Loess and Loess Geohazards in China

Loess and Loess Geohazards in China

Yanrong Li, Jingui Zhao and Bin Li
Taiyuan University of Technology, PR China

CRC Press
Taylor & Francis Group
Boca Raton London New York

CRC Press is an imprint of the
Taylor & Francis Group, an **informa** business

A BALKEMA BOOK

CRC Press
Taylor & Francis Group
6000 Broken Sound Parkway NW, Suite 300
Boca Raton, FL 33487-2742

First issued in paperback 2019

ISBN-13: 978-1-138-03863-9 (hbk)
ISBN-13: 978-0-367-88915-9 (pbk)

Typeset by MPS Limited, Chennai, India

Library of Congress Cataloging-in-Publication Data

Names: Li, Yanrong (Writer on geology), author. | Zhao, Jingui, author. | Li,
    Bin (Writer on geology), author.
Title: Loess and loess geohazards in China / Yanrong Li, Jingui Zhao & Bin
    Li, Taiyuan University of Technology, PR China.
Description: London : CRC Press/Balkema, [2017] | Includes bibliographical
    references and index.
Identifiers: LCCN 2017029842 (print) | LCCN 2017032733 (ebook) | ISBN
    9781315177281 (ebook) | ISBN 9781138038639 (hardcover : alk. paper)
Subjects: LCSH: Loess—China. | Sediments (Geology)—China. | Silt—China. |
    Sediment compaction.
Classification: LCC QE579 (ebook) | LCC QE579 .L5325 2017 (print) | DDC
    552.5—dc23
LC record available at https://lccn.loc.gov/2017029842

Published by: CRC Press/Balkema
                Schipholweg 107C, 2316 XC Leiden, The Netherlands
                e-mail: Pub.NL@taylorandfrancis.com
                www.crcpress.com – www.taylorandfrancis.com

Visit the Taylor & Francis Web site at
http://www.taylorandfrancis.com

and the CRC Press Web site at
http://www.crcpress.com

Table of contents

Preface

Loess covers approximately 6% of the Earth's land surface, mostly confined to the latitude range of the world's wheat belt. The Loess Plateau of China, which is home to a population of nearly three hundred million, has the thickest and most complete loess strata, where most frequent loess geohazards in the world occur. Even expanding human activities increase the frequency and scale of loess geohazards.

A systematic research on loess in China commenced, after the founding of People's Republic of China. Material composition, structure, fossil content, paleomagnetic records, spatial variation, and engineering characteristics of loess were all documented. Evidence of aeolian origin of surficial deposits in the Loess Plateau helped to understand Earth's climate and environments during the past 2,480,000 years since loess deposition started. The global paleoclimatic change during this time was corroborated by oxygen isotope studies on deep-sea drill cores and polar ice cores.

In recent years, the research on loess shifted to engineering geological characteristics and geological hazards mainly focusing on landslide and collapse, ground fissures, soil erosion, etc. Hence, recent literature discussing the unique properties of loess and loess geohazards from the perspectives of topography, geological history, regional and stratigraphic distribution, microstructure, and physical and mechanical properties are presently lacking. Recognizing the unique nature of loess and the extensive, serious, and far-reaching influence of loess geohazards on safety and productivity, this book focuses on loess and loess geohazards in China.

Loess is a Cenozoic loose and earthy aeolian deposit of yellow dust with a uniform porous structure. It has no bedding, is upright in structure, and rich in soluble salts. The mechanical properties of loess, such as permeability, shear strength, and tensile strength, are mainly determined by its structure and material composition. The structure and material composition are closely related to the genesis of loess and the geomorphic units. The regularity of loess geohazards is directly related to the unique physical and mechanical properties of loess and the special geological environment. Following these relations, we divided this book into eight parts to systematically explore the characteristics of loess and loess geohazards in China.

Chapter 1 introduces the origin and spatial distribution of loess both in China and in the world, emphasizing definition, evidence of aeolian origin (material composition, spatial zoning of grain size, source area, transportation mode), and process of formation and evolution.

Chapter 2 describes the features of loess landforms in China and their formation and evolution processes, starting with the types of primary landforms (landforms between gullies: platforms, ridges and hillocks), to gullies and secondary loess landforms (erosion landforms: dish, cratering, wall and column).

Chapter 3 elaborates on the loess microstructural characteristics (size and shape, existence form, contacts, cementation, particle arrangements and pores), classification of microstructure, particle size zoning across the Loess Plateau, microstructural characteristics under SEM, and the microstructural differences between the particle size zones.

Chapter 4, on the basis of the summary of the stratigraphic classification of loess (according to petrological characteristics, fossil content, erosion surface, etc.), describes the main physical and mechanical properties of loess (dry density, natural moisture content, plasticity index, liquidity index, compressibility coefficient, compression modulus, and collapsibility coefficient), the range of index values of physical and mechanical properties in each layer.

Chapter 5 introduces loess permeability, which exerts a significant effect on the reduction of strength and the occurrence of loess geohazards. It also examines the distributions of permeability in different strata and expounds on the main factors that affect permeability (dry density, buried depth, confining pressure, permeation time, freeze–thaw cycle, and anisotropy).

Chapter 6 introduces the loess shear strength, which plays an important role in the stability of loess and engineering applications. It examines the internal and external factors that affect the shear strength of loess, and the influences of main factors such as age, moisture content, compaction degree (dry density, void ratio), test methods, and shear direction.

Chapter 7 introduces the effects of tensile strength on loess geohazards and describes the parameters of tensile strength. The effects of main factors such as moisture content, dry density, size of specimen, tensile rates, consolidation pressure and compaction degree, and test methods on the tensile strength are discussed in detail in this part.

Chapter 8 elaborates on controlling factors, distribution law and failure modes of two main types of loess geohazards that are widely developed under different conditions: shallow collapse and deep-seated landslide.

This book presents a concise and up-to-date collection of research on loess and loess geohazards in China, and identifies the problems that need further study. We hope that this book can serve as a principal reference for researches infected in loess and loess geohazards in China and around the world. The structure of this book makes it also suitable for undergraduate and postgraduate students of engineering geology.

This book would not have been possible without a large number of research results on loess across China. We thank Professor Adnan Aydin from the University of Mississippi (USA) and Professor Yongxin Xu from the University of Western Cape (South Africa) for reading various versions of this book. We also thank a number of graduate students who aided in the data collection and preparation of the charts.

Yanrong Li
Professor of Earth Sciences and Engineering
Taiyuan University of Technology, China

Foreword

With the rapid development of economies, countries around the world are now paying more and more attention to environmental problems. Geological environment is the basis of ecological environment; moreover, both are interdependent and mutually influence and restrict each other, forming a relatively unified whole closely related to human activities.

Although the scope of human activities is confined to the shallow surface of the solid lithosphere, the breadth and frequency of the impact from human activities on the ecological and geological environment are now expanding significantly. If these activities cannot be reasonably regulated and controlled, the original balance of environment may be disturbed, eventually leading to the imbalance and catastrophic environmental disasters. However, the current theories and management experience cannot fully solve the geological environment problems today as they are relatively complex. We, thus, still face severe challenges in disaster prevention and protection of geological environment.

Loess covered areas form a special geological environment, which features wide distribution and complex surface morphology and is prone to serious soil erosion and frequent geohazards. Hence, the geological environment and ecological environment of loess covered areas are extremely fragile and are particularly interdependent. With the implementation of the "West Development Strategy" and the "Silk Road Economic Belt" and the "21st Century Maritime Silk Road Initiative", human activities in the Loess Plateau of China increased in intensity, thus expanding influence on the fragile eco-geological environment balance in loess areas. Therefore, the coordinated and sustainable development of eco-geological environment and human activities in loess areas is more typical, more prominent, and more complex than that in other regions.

To effectively study and solve the aforementioned problems, it is necessary to obtain insight into the characteristics of the geological environment and the engineering characteristics of loess. Professor Yanrong Li, Dr. Jingui Zhao, and Dr. Bin Li chose Chinese loess and loess geohazards as their research subject and expounded on the characteristics of the geo-environment of loess areas, as well as the main patterns and distribution rules of loess geohazards in China, from the aspects of its origin, distribution, morphology, physical and mechanical properties, and geological disasters. Such work highlights the concern of the engineering geology society in China about the economic construction, environmental protection, and sustainable development in loess areas, as well as their enthusiasm to participate in solving the above problems.

This book integrates the wisdom of Chinese scientific and technical workers engaged in the research or productive practice of loess geology, and presents the latest research achievements from different aspects in this field in China. To a certain extent, it represents the frontier of loess engineering geology in China. Emphasis must be placed on the young editors of this book, who represent the young generation of Chinese engineering researchers with strong academic advocacy and desire to pursue scientific truth. I fully believe that the publication of "Loess and Loess Geohazards in China" in English will positively promote the economic and environmental sustainable development in the central and western regions of China, and improve the overall understanding of loess and loess geohazards in China.

Runqiu Huang
Professor of Engineering Geology
State Key Laboratory of Geohazard Prevention
and Geo-environment Protection, China

Foreword

The loess strata in the Loess Plateau of China hold abundant information about the global natural environment and climate changes, as well as the occurrences of surface disasters in the past 2.4 million years. Moreover, nearly a third of China's geohazards occur in the Loess Plateau every year. The study of loess has been of interest to international geoscience community, with the origin of loess and the environmental changes in the Loess Plateau the two always hot research topics. Scientists of the older generation, such as Zhongjian Yang, Bingwei Huang, Tungsheng Liu, Xianmo Zhu, Deqi Jiang, Zonghu Zhang and Zhisheng An have conducted in-depth and systematic scientific research that currently serves as our scientific theoretical basis for our study of the Loess Plateau.

With the implementation of the "West Developments Strategy" and the "Silk Road Economic Belt and the 21st Century Maritime Silk Road", the resulting large-scale engineering constructions have become an important cause that induces geohazards, particularly in loess areas. All of the major road engineering and water conservancy projects under construction or under consideration require large-scale slope excavation and fill subgrade, leading to disturbances to the geological environment. These disturbances, combined with the superposition of changes in the water environment, constantly induce new loess geohazards that in turn endanger the construction and operation of major projects.

Professor Yanrong Li, Dr. Jingui Zhao and Dr. Bin Li have scientifically classified voluminous documents and data related to loess studies in China in recent years. They carried out statistical studies and systematic elaboration of topics such as the origin and distribution of loess, loess landforms in China, loess microstructure, physical and mechanical properties of loess and loess geohazards in China. Using concise charts, they revealed the synergistic control law of loess geological structure and mechanical behavior of loess medium in the development of geohazards. They also represent the present situation of the research on loess and loess geohazards in China. This book can provide a platform for engineering geology academia to facilitate a systematic understanding of the engineering geological characteristics of loess. This work will play an important role in promoting the further study of loess geohazard modes, evolution mechanism of major loess hazards, and development law under complex dynamic conditions. It will likewise aid the development of a theoretical and technical method for warning and risk control of major loess geohazards.

Jianbing Peng
Professor of Engineering Geology
Chang'an University, China

Acknowledgements

The journey of pursuing this book would not have been possible without the perpetual support and encouragement of Professor Runqiu Huang and Professor Qiang Xu from the State Key Laboratory of Geohazard Prevention and Geo-environment Protection, China, Professor Jianbing Peng from Chang'an University, China, Professor Adnan Aydin from the University of Mississippi, USA and Professor Yongxin Xu from University of the Western Cape, South Africa. The authors would like to express extreme gratitude to them for their invaluable suggestions and critical and stimulating discussions on the subject. Many thanks are expressed to my colleagues and students at the Institution of Geoenvironment and Geohazards (GEGHI) of Taiyuan University of Technology, China. The permissions from the copyright holders for figure reproductions are also greatly appreciated.

About the Authors

Prof. **Yanrong Li**, PhD, works in the Department of Earth Sciences and Engineering at Taiyuan University of Technology, China. He is a funding Director of the Collaborative Innovation Center for Geohazard Process and Prevention at this university. He worked for ten years as a practicing engineer and field geologist in Hong Kong, Australia and China before he turned to be a full professor. The research of Prof. Li focuses on Quaternary Geology, Geoenvironment and Geohazards. He is a Chartered Geologist (CGeol) of the Geological Society of London and a Professional Member (MIMMM) of the Institute of Materials, Minerals and Mining. He served as an Associate Editor of KSCE – Journal of Civil Engineering and is now an Editorial Board Member of the Bulletin of Engineering Geology and the Environment. Professor Yanrong Li (li.dennis@hotmail.com) welcomes discussions on the contents of this book and diverse collaboration with experts of similar interests.

Dr. **Jingui Zhao**, has been in the Department of Earth Sciences and Engineering at Taiyuan University of Technology, China since September 2004. He obtained academic degrees of BSc (in Engineering Geology, 1998), MSc (in Mineral Resource Prospecting and Exploration, 2004) and PhD (in Tectonic Geomorphology, 2014) all from Taiyuan University of Technology. Dr. Zhao has great research interests in Structure Geology, Quaternary Geology and Geomorphology, and Geohazard. He has conducted research projects with outstanding outcomes, and published peer-review academic papers in these fields.

Dr. **Bin Li,** works in the Department of Earth Sciences and Engineering at Taiyuan University of Technology, China since 2016. He lived in the University of Bergen (UiB), Norway, for seven years, where he completed his MSc in Quaternary Geology and Paleoclimate in 2007, and PhD in Seismotectonics and Seismic Hazard in 2015. Prior to his PhD study in UiB, Dr. Li worked for three years as an engineer at the Earthquake Administration of Shanxi Province, CEA, China. His research focuses on Quaternary Geology, Earthquake Seismology, and Earthquake-induced Geohazard. He is involved in many research and has published a number of peer-review papers.

Chapter 1

Origin and spatial distribution

In Chinese, loess is called Huangtu, which means "yellow earth". Loess (Huangtu), which comprises newly formed sediments, covers approximately 6% of the Earth's surface area. The knowledge of the origin and distribution of loess is prerequisite to the study of loess landform, structure, physical and mechanical properties, and geohazards.

1.1 INTRODUCTION

The debate on the origin of loess in the last 100 years has generated over 140 different hypotheses, which cover almost all types of depositional processes, including alluvial, marine, lacustrine and colluvial, in-situ soil, extraterrestrial, glacial, and aeolian origins, etc. In 1834, Charles Lyell thought that loess in Western Europe was formed by flooding over river banks by comparing the loess with Nile river sediments and then hypothesizing the alluvial origin of loess. In 1841, Charpentier contended that loess is a very fine glacial clay deposit that resulted from the melting of glaciers and later hypothesized a glacial origin. In 1857, Beningsen–Forder proposed a hypothesis of marine origin, but it could not be accepted because of the absence of marine fossils and marine sedimentary structure in loess. In 1967, Pumpelly hypothesized a lacustrine origin in a study of the loess in northern China, but he himself rejected the hypothesis due to lack of evidence. In 1868, Ferri Gail put forward the hypothesis of a proluvial origin on the basis of observations that the structure and composition of loess are characteristic of formations caused by oxidization under atmospheric conditions and that loess landforms occur in the form of proluvial deposits that contain sand, gravel lenses, and rock fragments, which are derived from nearby mountain slopes and contain continental fossils. In 1916, Berg stated that loess is a product of moraine, alluvial, and slope wash sediments caused by weathering and pedogenesis during interglacial periods. In 1920, considering the homogeneity of loess with vast spatial distributions, Keilhack proposed an extraterrestrial origin. In 1857, Virlet D'Aoust first proposed the hypothesis of an aeolian origin by considering that the loess in the Mexico plateau is composed of cyclone dust. In 1882, Richthofen further elaborated the aeolian origin of loess based on the composition and structure of loess uniformity and slight horizontal bedding structures regardless of altitude and other evidence (as discussed later) (Li & Sun, 2005). In 1965, Liu described the loess system systematically, and thereafter, the aeolian origin of loess was finally recognized universally.

1.2 DEFINITION OF LOESS

Combining the research results on loess deposits across China, Liu (1985) divided loess deposits into primary and secondary (loess-like sediments). Primary loess is a wind-blown deposit without secondary disturbances and bedding planes. It mainly comprises yellow silt particles with high porosity and is rich in carbonates. The loess distributed in Shanxi, Shaanxi, Gansu, and other parts of the Loess Plateau is representative of primary loess. Secondary loess (loess-like sediments) is formed by the re-deposition of primary loess in arid and semi-arid regions. Secondary loess can be found in different depositional environments in addition to aeolian deposits. It is a yellow, powder-like bedded sediment that contains layers of silt, sand, and gravel (Liu, 1985). Primary and secondary loess differ in distribution, occurrence, topographic and geomorphic features, thickness, structure, microstructure, uniformity, particle composition, lithology, mineral composition, and collapse potential, all of which point to different loess origins (Table 1.1).

Sometimes, secondary loess and primary loess look similar in field, hence the tendency to confuse the two types. The distinctions between them (Table 1.1) help clarify the origin of loess deposits, and the identification of these distinctions is a milestone in the study of loess geology. After deposition, the macroscopic characteristics of secondary loess and their engineering geological characteristics become similar to those of primary loess. Special attention should be paid to the identification of the origin of loess deposits under investigation.

1.3 EVIDENCE OF AEOLIAN ORIGIN OF LOESS

In addition to the characteristics listed in Table 1.1, loess of aeolian origin shows a distinct material composition, spatial particle distribution, source relation of its materials, and features related to its mode of transport.

1.3.1 Material composition

Theoretically, the dynamic geological processes (such as alluviation, colluviation, and proluviation) of the source during transport and after deposition lead to differences in the material composition of sediments. The mineral and chemical compositions and other components of primary loess are almost identical in space; thus, hypotheses other than that related to the aeolian origin of loess are difficult to support.

Colloidal particles dispersed in loess in different locations are mainly composed of illite, quartz, kaolinite, montmorillonite, goethite, hydrogoethite, and pyrophyllite (Table 1.2). All these particles show that the provenance of loess in China is consistent.

The chemical composition and content of the loess in different regions along the middle reaches of the Yellow River is relatively uniform. For example, SiO_2, Al_2O_3, CaO, Fe_2O_3, and MgO account for about 54.06%, 11.36%, 7.74%, 4.42%, and 3.91% of the total content (Fig. 1.1). The high oxide content indicates that the source materials are formed in an oxidation environment, whereas the presence of soluble salt content indicates an arid environment.

Table 1.1 Comparison of characteristics of primary and secondary loess deposits in China (Li & Sun, 2005).

Characteristics	Primary	Secondary
Distribution and occurrence	Continuous distribution of thick layers. Covers basins, slopes, hills, erosion surfaces, valleys, and terraces. Often in contact with bedrock.	Forms strips, sheets, and scattered patches in front of alluvial fans, low terraces, and alluvial plains (occasionally on hillsides as small patches). Often in contact and alternate with loose sediments.
Topography and geomorphology	Often forms platforms (Huangtu Yuan in Chinese), ridges (Huangtu Liang), and hillocks (Huangtu Mao). Mostly shows a wavy, undulating terrain. Present day topographic relief is consistent with the underlying terrain.	Often seen on plains and alluvial plains or alluvial terraces. The ground is generally flat, and is less affected by the paleo-landform.
Thickness	In general, a few meters to 200 meters.	Usually a few meters to about 10 meters.
Color	Grayish yellow or brownish yellow. Consistent throughout the region and profile.	Yellowish gray, pale brown, or dark brown.
Organization structure	Loose, brittle, uniform, and porous. Not bedded.	Hard, brittle, uneven, bedded. Includes micro inclusions and a few large pores.
Field feature	Layers of old paleo-soil and calcareous nodules. Columnar joint development. Nearly vertical cliffs.	Layers without paleo-soil. Columnar joint is not developed. Does not usually form vertical cliffs.
Uniformity	All layers are uniform in particle size, minerals, and chemical components vertically and horizontally. No interlayers of sand and gravel (except for deposits near deserts).	Layers are heterogeneous and often interlayered with sand and gravel. Particle size, minerals, and chemical composition change significantly over large areas.
Particle size and distribution	Weight content of silt (0.005–0.05 mm) >50%. Particles >0.25 and <0.005 mm are few. Well sorted.	Weight content of silt <50%. Particles >0.25 mm are sometimes very high in content. Particles <0.005 mm are sometimes dominant. Poorly sorted.
Lithofacies change	Away from the desert, particles gradually become fine, and Al_2O_3 and Fe_2O_3 contents gradually increase.	Away from the mountain, particles become fine, and the mineral and chemical composition changes significantly.
Mineral composition	Quartz and feldspar are the main minerals. Also contains a large amount of unstable minerals. Minerals are weakly weathered and similar over large areas and are not related to nearby mountain lithologies or underlying bedrocks.	Mainly composed of quartz and feldspar; the unstable mineral content is low. The weathering of minerals is advanced, and mineral composition is related to the bedrock. Regular changes occur in the mineral composition of watersheds and river sources.
Collapsibility	Collapsible. Easily produces ground sink and piping.	Wet trapping is small. Does not easily produce ground sink. Piping is not common.
Origin	Aeolian	Alluvial, colluvial, proluvial, etc.

Table 1.2 Mineral composition of loess in the middle reaches of the Yellow River (Liu, 1964).

Location	Mineralogy of colloidal particles
Lanzhou, Gansu	Illite, quartz, montmorillonite, kaolinite, organic matter, white mica
Yuzhong, Gansu	Illite, quartz, kaolinite, montmorillonite, goethite, hydrogoethite, pyrophyllite
Tianshui, Gansu	Illite, quartz, montmorillonite, kaolinite, goethite, hydrogoethite
Pingliang, Gansu	Illite, quartz, montmorillonite, kaolinite, goethite, hydrogoethite, pyrophyllite
Huanxian, Gansu	Illite, quartz, montmorillonite, kaolinite, goethite, hydrogoethite, pyrophyllite
Longdong, Gansu	Illite, quartz, kaolinite, montmorillonite, beidellite, goethite, hydrogoethite, glauconite, chlorite

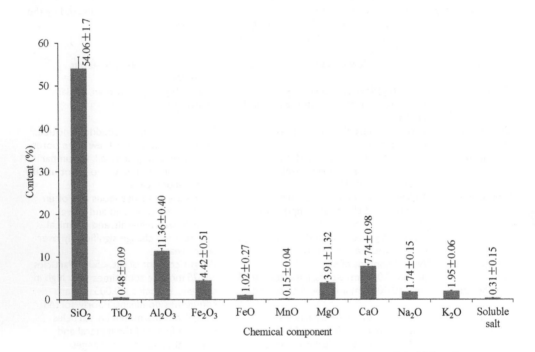

Figure 1.1 Chemical composition of loess along the middle reaches of the Yellow River.

The fossils found in loess are mainly composed of vertebrates, snail, pollen, etc. The vertebrate fossils belong to 56 families and 95 species, among them herbivores accounted for 82.6%, and carnivores accounted only for 15.4%, most of which are wolves, foxes, badgers, jackals, and other small animals, and large carnivores are few. In the middle and late Pleistocene loess layer, gastropods include as many as 29 species of land snail fossils, such as cathaica, metodonia, and pupilla. Pollens are mainly from dozens of plants, such as pine, spruce, fir, hemlock, oak, and compositae. The occurrence of these fossils indicates that loess was deposited in a typical arid grassland environment.

The minerals of the coarse grains of loess are mainly quartz and feldspar, with the average chemical and mineral compositions being similar to those of granite or granodiorite, which represent the average composition of the Earth's crust. The chemical and mineral compositions of loess do not change significantly across vast lands and over time, and they do not show any relation to the local bedrock. These characteristics indicate that the only possible type of energy that can transport and uniformly mix different raw materials across vast regions is wind.

1.3.2 Spatial zoning of grain size

In the Loess Plateau of China, the grain size of the loess in the same age appears to become increasingly fine from the northwest to the southeast, and the loess can be separated into three belts, namely, the sandy loess, silty loess, and clayey loess belts (Fig. 1.2). Figure 1.2 shows the locations of the loess studies listed in Table 1.2. The region to the northwest of Xingxian, Jiaxian, Mizhi, and Tongxin is the sandy loess belt with a mean particle size of 4.939 ± 0.203 Φ. The region to the southeast of Taiyuan, Wucheng, Pingliang, Yuzhong, and Huangling is the clayey loess zone with a particle size of 6.049 ± 0.211 Φ. Between these belts is the silty loess zone (Fig. 1.2) with a particle size of 5.611 ± 0.308 Φ. The distribution patterns suggest that the material source areas of the loess may be located to the northwest of the Loess Plateau, including the Tengger Desert, Mu Us Desert, and Hetao area. The patterns also indicate that the loess was transported from these areas by the northwest wind.

1.3.3 Materials of the source area

The grain size composition and calcium carbonate content of materials in source areas are similar to those of the Loess Plateau. Coarse sand (0.01–0.05 mm) accounts for 8%–23%, fine silty sand (0.005–0.01 mm) accounts for 1%–10%, coarse clay (0.001–0.005 mm) accounts for 2%–20%, and clay (<0.001 mm) accounts for 5%–23% (Nanjing Soil Research Institute, 1978), respectively. Calcium carbonate content is between 8% and 30%. The results of grain size distribution and chemical composition analyses show that the parent material consists of silt, clay, and calcium carbonate and that it was formed by biological, physical, and chemical processes in the soil horizon. The semi-arid, arid and desert steppe environments can provide a large amount of silt, clay, calcium carbonate, and other substances (Fig. 1.3). These areas (about 50–400 km^2) are bounded by the Kent and Tangnushan Mountains to the north, the Da Hinggan Mountains to the east, the Great Wall and Kunlun Mountains to the south, and the Tianshan Mountain and Pamir Plateau to the west (Zhu, 1983; Liu, 1985).

1.3.4 Transport mode

Studies on atmospheric circulation associated with cyclones and anticyclones, and even small-scale local vortices, show that the cold air at the middle and high latitudes and the high pressure are important factors that lead to dust storms and that the strong northwest wind is the main force behind sand and dust transport. Sand and dust, which are lifted by the air from the dry ground during the upward flow of a cyclone, are transported for long distances. The westerly wind belt over the inland arid region of central Asia and the Loess Plateau exerts a significant effect on the long and distinct transport of atmospheric dust. China is located in the East Asian monsoon region,

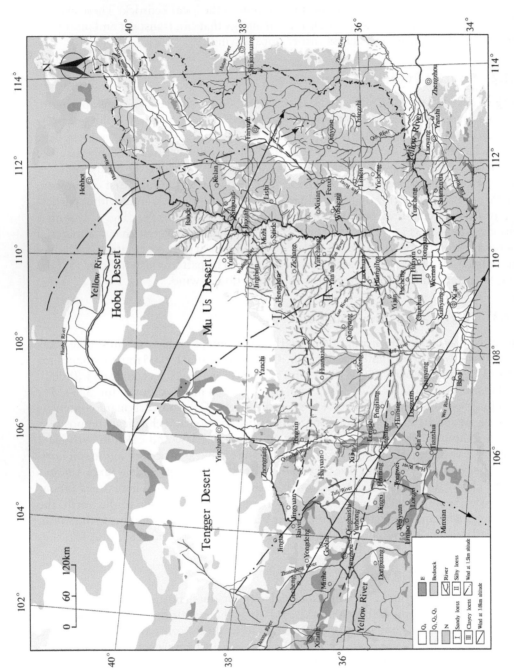

Figure 1.2 Zonation and particle size of loess in the middle reaches of the Yellow River (Modified from Liu, 1985). Curved dash lines are the boundaries between the sandy loess, silty loess, and clayey loess belts; solid and dash lines with arrows indicate transport direction of sources.

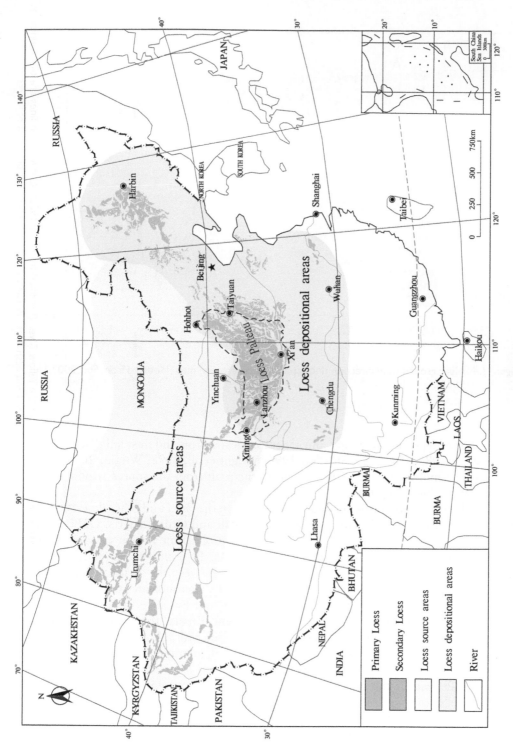

Figure 1.3 Loess sources and depositional areas in China (Modified from Li & Shi, 2017).

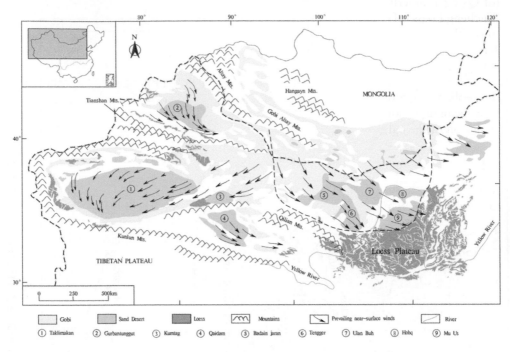

Figure 1.4 Near ground wind direction (indicated by small arrows) map (Modified from Sun, 2002; Liu, 1965).

where the wind prevails in winter and generates dust storms and dust fall (Richthofen, 1982; Zhu, 1983; Liu, 1985; Smalley, 1996; Wright et al., 1998; Wright, 2001). The near surface monsoon wind direction shows the path from the source region to the areas of deposition (Fig. 1.4).

Wind erosion landforms are widely developed in the Gobi and other deserts to the northwest of the Loess Plateau, such as the Alashan and Ordos deserts. Only coarse materials remain on the ground surface. The direction of movement of sand dunes and sand ridges in these areas points to loess areas. By contrast, fine particles are the main materials in loess areas. The clear contrast between the materials (especially particle size) of the northwest areas and the Loess Plateau indicates that wind is the main energy that transports fine materials from the northwest to the Loess Plateau. In typical loess regions, such as Gansu of China and Russia, the roundness of loess particles (0.03–0.04 mm) is less than that expected for hydraulic transport (normally above 0.15 mm). This characteristic supports the aeolian origin of loess.

The thickness of loess appears to get thinner from northwest to southeast over the Loess Plateau. If the wind is locally weakened by mountains, the loess on the windward slopes is generally with greater thickness and is located at higher elevation than that on the leeward slopes. Loess continuously, like blanket, covers a variety of undulating terrains. The Malan Loess (of Q_3 age) continuously distributes over the first and second terraces, and it sometimes covers highly-elevated ridges. The Lishi Loess (of Q_2 age) underlies the Malan Loess and covers the third and fourth terraces. The Wucheng Loess (of Q_1 age) is deposited on the fifth terrace. Along the valleys, it is easy to see the different layers of these loess deposits stacked together. This phenomenon can also be observed in the northeast China and as far away as Ukraine. This type of continuous layering can only be explained by the aeolian origin of loess.

From Richthofen in 1882 to Liu in 1965, many researchers have come to recognize the aeolian origin of loess. The formation of loess is closely related to the large-scale East Asian monsoon system (Fig. 1.4), which also makes loess a good candidate in the evolution of the paleoclimate since the Quaternary.

1.3.5 Process of formation and evolution

After defining the material production, transportation and deposition mechanisms, Liu (1985) depicted the process of formation and evolution of the loess (Fig. 1.5):

1   Production of material – in the arid desert of Gobi and the bordering mountain ranges, a large amount of fine particles of SiO_2, Al_2O_3, CaO, etc. are produced due to weathering and glacial movement;
2   Transportation – sand and dust are transported by the cyclone and anticyclone;
3   Deposition – as the air mass moves, the sand and dust subside and accumulate in the arid and semi-arid environments;
4   Formation – sand and dust deposits are bonded by a newly developing carbonate matrix and create the structure of loess;
5   Evolution – the loess deposit will be subjected to different evolution style depending on the specific covering and stress conditions of it. If the loess is deeply buried by another deposit, it may be compacted to form rock layer, such as silty sandstone. If the loess is transported a long distance and redeposited to the humid and semi-arid areas, then other types of sediments (alluvial deposits, for instant) will form. If the loess is transported only a short distance and redeposited to the dry and semi-arid areas, then it will form the secondary loess; if the loess is weathered in-situ, then it will form a pedogenic soil; if the in-situ weathered soil is covered by later sediments, then it will form a buried paleo-soil; if no new sedimentation occurs and the weathering is prolonged in-situ, then it will form a weathering crust; if the weathering crust is covered by the later sediments, then it will form the buried weathering crust.

1.4 SPATIAL DISTRIBUTION

1.4.1 Distribution of loess in the world

Primary and secondary loess deposits are intermittently distributed in arid and semi-arid regions in the mid-latitudes of northern and southern hemispheres, including inland temperate deserts, semi desert edges, periphery of Quaternary glacial ice sheets and some coastal zones. The total loess area of 13,000,000 km² (estimated, includes patchy loess deposits), occupies about 9.8% of the Earth's land surface (Li & Sun, 2005). Loess can be globally divided into "warm" area loess connected with deserts and "cold" area loess connected with continental ice sheets (Обрлев, 1958).

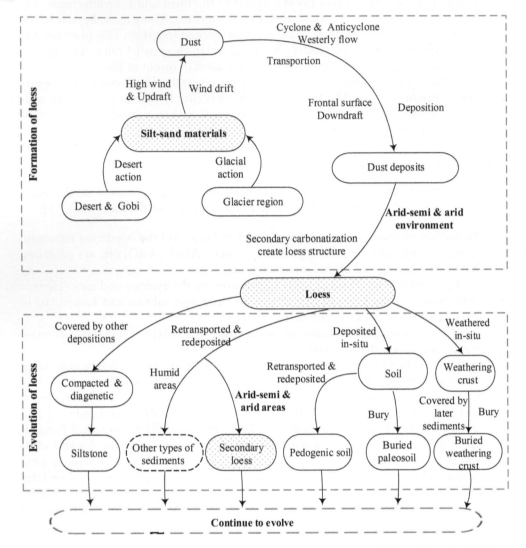

Figure 1.5 Schematic diagram of the formation and evolution process (Modified from Liu, 1985).

In Europe, North America and Siberia, loess is mainly distributed adjacent to the peripheries of Quaternary continental ice sheets, while the loess in Central Asia, China and South America mainly distributed in the peripheries of deserts. It is shown that loess formation is related to the anticyclonic winds generated by the high pressure center of continental ice sheets and deserts.

As shown in Figure 1.6, large tracts of loess cover Europe, Asia, North America and South America, and smaller loess bodies are found covering parts of Africa, Arabian Peninsula, New Zealand and Australia. Limited to the different details of current

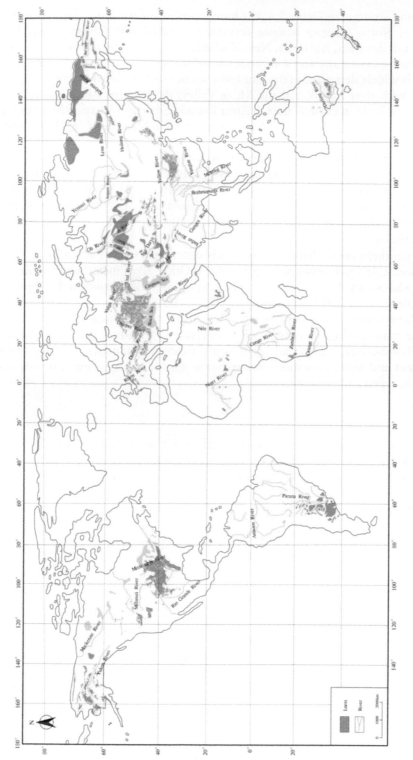

Figure 1.6 Distribution of loess on the Earth's surface (Modified from Li & Shi, 2017).

literature and sources, in Figure 1.6, the loess distribution in China, Europe and mid-continent of North America is more accurate compared with Siberia, Central Asia, Alaska, South America, Australia, New Zealand, Africa and Arabia Peninsula, where the boundaries, continuity and soil properties of loess distribution are less precise.

Loess is widely deposited in the mid-latitudes of northern hemisphere, particularly associated with river basins, e.g. the Rhine Valley, the Danube River Basin, and the Dnieper River Basin in Europe, the Ob River Basin in Siberia of Russia, the Amu Darya and Syr Darya Basins in Central Asia, the reaches of Yellow River in China, and the Mississippi River Basin in America.

Asia is the most widely loess distribution area in the world, with a latitude range between and 32°N (Nanjing, China) and 74°N (Bol'shoy Lyakhovskiy, Ostrov Island, Russia). The loess in Europe is distributed between 40°N and 62°N in latitude. In the mid-continent of North America, loess area is located between 29°N and 49°N, and in North Africa, Arabia Peninsula, loess is sporadically distributed, between 10°N and 35°N in latitude.

In the southern hemisphere, loess is mainly distributed in the Parana River Basin of South America, latitude between 20°S and 40°S. Other areas including New Zealand, Australia and South Africa, loess exhibits a sporadic distribution and is limited to middle latitudes (Liu, 1985; Li & Sun, 2005).

The Earth's surface where is covered by primary and secondary loess deposits, is one of the main geological environments necessary for human survival and development. Loess-covered areas are also home to dense industrial and agricultural development and high population concentration. Such areas are significant for wheat production, especially in China.

1) Europe

Loess covers approximately 1/5 of the total surface of Europe, especially in the foothills of the Alps and a low mountain range belt in the north, the Danube basin and other various river basins, and the lowlands in eastern Europe (Haase et al., 2007; Muhs et al., 2014; Jefferson et al., 2003).

European loess deposits are distributed in three main zones from west to east.

a)   Eastward to longitude 15°E, loess is found mostly (and is thickest) in a latitudinal band at about latitude 50°N, between the edge of the former continental Scandinavian ice sheet in the last glacial period to the north (today's northern Germany and northern Poland) and the Alpine glaciers in the south.

In North-Western Europe there is a continuous loess belt which mainly consists of pure loess and loess derivates. Its width varies from 10 to 200 km (between Normandy and Champagne, Northern France). And the loess there is distributed in areas situated above 300 m a.s.l. In major parts of Central Europe from the Harz Mountains (Northern Germany) to the northern fringe of the Carpathians, a continuous loess belt of 50–100 km wide was formed. Some 'fragmentary loess cover' presents in regions at altitude of 400–600 m. In regions higher than 600 m, loess derivates occur on plateaus surrounded by mountain chains. Good examples of such a zoning are the fringe of the Paris basin, the Ardennes, the northern border of the Ore Mountains (Germany), and the Alpine Mountains.

b)   Between longitude 15°E and 25°E, the thickest loess sequences are mostly located south of the Carpathian arch.

At lower latitudes, thick loess deposits have been described in the Middle Danubian Plain along the Danube River, mainly including Austria, Slovakia, Hungary, Croatia and Serbia.

c)   Eastward from longitude 25°E, most of the European loess region is in Eastern Europe, corresponding to the wide Ukrainian and Russian plains. It might be defined as lying within latitude 44–56°N, and longitude 24–48°E, now comprising Ukraine, Belarus, Moldova and parts of south-eastern Russia.

It is limited in the south of Eastern Europe by the Black Sea and the Caucasian Mountains and in the north-east by the Smolensk-Moscow Upland. In these regions there are patchy loess-like sediments of different types which do not create such clear zones as Central and Western Europe. The Eastern European loess cover can be divided into several regions according to different facies or thickness. Large loess areas are located in the basins of south of Eastern Europe, especially the Lower Danubian Plain. According to current knowledge, the thickest loess covers recorded in Europe may be found here. In Northern Bulgaria, loess sediments are more than 100 m thick. This area is connected (to east) with the loess areas of the Eastern European lowlands. Around the coast of the Black Sea, the loess is more than 10 to 20 m over large areas.

Finally, outside the above areas of continuous loess occurrence, discontinuous loess sediments occur in the surroundings of the Po-plains and in the valleys of the Rhone and the Garonne (France). Moreover, numerous loess patches are known in Spain, Central and Southern Italy and in the Southern Balkan Peninsula.

Much of the loess cover in Eastern and Central-Eastern Europe has been redeposited by the Danube River. Other areas of loess are associated with rivers such as the Seine, Somme, and Rhone in France, the Rhine in Western Germany, the Po in Italy, and the Dnieper in Russia, Belarus and Ukraine.

2) Asia

Loess mantles extensive regions in Asia, especially in the largest mid latitude arid–semiarid zone in the Northern Hemisphere (Muhs et al., 2014; Jefferson et al., 2003; Murton et al., 2015; Péwé & Journaux, 1983; Crouvi et al., 2010).

In Central Asia and China, most loess bodies are bordered by or in the downwind areas of gobi (stony desert) and sand deserts. In this part, including Central Asia and China, four main loess zones in Asia are described as follows.

a)   The most widespread loess deposits in Asia occur in China, centered in the Loess Plateau, and will be described in later chapter in detail.

b)   In Central Asia, loess covered areas are located in Tajikistan, Kyrgyzstan, Turkmenistan, Uzbekistan, and Kazakhstan. Central Asia have medium-sized dispersed loess patches; the loess area is large and this is a very important loess area given the close association with major cities such as Tashkent, Alma-Ata and Dushanbe.

In Central Asia, loess deposits are adjacent to mountain regions and dominantly cover piedmonts and hills. In contrast to the Chinese Loess Plateau, loess in Central Asia mostly accumulates on the windward slopes of the Central Asian orogenic belt (including the Tian Shan, Kunlun, Hindu Kush, and Pamir Mountains), where loess can be found at elevations of up to 2,500–3,000 m. Generally, loess deposits in Central Asia are several tens of meters thick, except in certain regions, such as Tajikistan, or in the vicinity of Tashkent, where the loess strata can be up to 100–200 m. Two main rivers, the Syr-Darya and the Amu-Darya, they were substantial movers of loess material, out from the Tian Shan and towards the Kyzyl Kum desert and the Aral Sea.

c)   In Siberia of Russia (east of the Ural Mountains), there are multiple distribution of loess; we propose to describe in 5 parts: Orsk–Omsk (Western Siberia), Tomsk-Barnaul, Krasnoyarsk-Kansk, Irkutsk, Northeastern Siberia.

Orsk–Omsk (Western Siberia): Loess deposits all down the Ural River from Orsk (to south). In north and east of Orsk a large loess deposit is indicated but this belongs to the north-flowing Tobol River; the northern limit approximately by the town of Kurgan. The east of this region stretches towards Omsk and the Irtysh River. An extensive loess deposit all lie along the Irtysh River in south of Omsk. The Irtysh delivers material directly from High Asia to the eastern part of this zone. The river Ishim supplies material to the middle part of the zone; it flows on to join the Irtysh, as does the Tobol in the west.

Tomsk-Barnaul: Associated with river channels flowing to consolidate into the Ob River, loess deposits stretch south from Tomsk. Barnaul is a useful locator of this area. Loess material comes from the south and rivers flow to north.

Krasnoyarsk-Kansk: This loess region is associated with the Yenisei River. The town of Kansk also serves as a locator. Yeniseysk represents the northern limit. Loess material comes from the mountains of the south, and mostly is carried by the large north-flowing rivers.

Irkutsk: The deposits near Irkutsk lie along the Angara River; they appear as a relatively small outlier of collapsing ground to the north of Baikal Lake. Its geographical position suggests a particle origin in the mountains of the south, with major transportation by the Angara River. Lake Baikal possibly has an important role to play; there are 300 streams feeding into Baikal, but only one outlet – the Angara River. Baikal may be an intermediate source of loess material.

Northeastern Siberia: A special kind of loess in some areas of northeastern Siberia, central and northern Alaska and the Klondike region of Yukon (Canada) is often called 'Yedoma' or 'Ice Complex' in both the Russian and North American literature. Yedoma refers to encompassing distinctive ice-rich silts and silty sand penetrated by large ice wedges, resulting from sedimentation and syngenetic freezing during the late Pleistocene. Yedoma can rich in organic carbon. In northeastern Siberia, loess (yedoma) is mainly distributed in the Central Yakutia lowland and Kolyma Lowland.

In the central Yakutian lowland– directly west of western Okhotsk Sea– silty yedoma deposits occur in the surface of where continuous permafrost is about 400–700 m thick. Yedoma mantles upland terraces and low plateaus throughout unglaciated south-central Yakutia. In general, silt is thickest in the bottoms of small valleys and on lower slopes and thinner on the highlands. Yedoma is thickest along the south side of the lower Aldan River valley and the east side of the Lena River valley. On the Tyungyulyu Terrace it reaches a maximum thickness of 60 m near Syrdakh (about 70 km northeast of Yakutsk) in the general region between the junction of the Aldan and Lena Rivers. Along the west side edge of the Lena River valley, silt is 10–25 m thick.

The Kolyma Lowland forms the easternmost segment of the northeast Siberian coastal plain – comprising the Yana, Indigirka and Kolyma lowlands – to the south of the East Siberian Sea. Yedoma is widespread in the Kolyma Lowland. Its southern limit in this region stretches along the margins of river floodplains, where the front of adjacent uplands. Such yedoma is metres to tens of metres thick. The depositional land surface that formed along the top is a flattish 'yedoma surface' that was subsequently modified by thermokarst activity. The yedoma surface is well developed between the Omolon and Bol'shoy Anyuy rivers, where it is termed the Omolon-Anyuy yedoma ($>1000\,km^2$). Yedoma also occurs in some low mountainous areas south and west of the Kolyma Lowland.

d)   Loess is also sporadically reported in Japan, northern India, northeastern Iran, but lacks of loess distribution maps of these areas in current.

For example, Tanino et al. (2015) described sporadic Japanese loam (loess) in four areas, Esashi, Shiriyazaki, Isozaki-Ajigaura, and Kashima, of northern and eastern Japan. These loess dunes mainly overlie the top of rock coastal cliff or terrace.

3) North America

In North America, loess is found mostly beyond (to south) the margins of where the Laurentide and Cordilleran ice sheets and mountain glaciers advanced during the last glacial period (Today's Michigan and Minnesota of USA), and southward reach to Louisiana of USA (Muhs et al., 2014; Iii et al., 2003; Murton et al., 2015).

The loess area in North America can be divided into 4 parts: Mid-continent region, Colorado Plateau region, and Columbia Plateau region in the conterminous United States of America, then the Alaska (USA) and Klondike of Yukon (Canada).

a)   Mid-continent region

The greatest extent of loess is found in the mid-continent region of North America, including the greater Mississippi River drainage basin and the Great Plains as well as the source areas of the Mississippi River Basin's loess sequence.

Most of the loess strata in such area are composed of Peoria Loess. The Peoria Loess (also known as Peoria Silt, Peoria Formation, Peorian loess, or Wisconsin loess), deposited during the Last Glacial maximum (LGM) in North America, is probably the thickest LGM loess in the world. Peoria Loess is more than 48 m thick at Bignell Hill in central Nebraska, 41 m thick at the Loveland paratype locality in western Iowa, 10–20 m thick at many sections along the Mississippi Valley and 10 m thick along the lower Wabash Valley in southeastern Indiana. Even in eastern Colorado, distant from the Laurentide Ice Sheet, Peoria Loess is as thick as 10 m.

b)   Colorado Plateau region

There are extensive aeolian silts, sandy silts and sandy clay loams on upland bedrock areas of the Colorado Plateau, and an extensive area of thin loess in southwestern Colorado and southeastern Utah.

The range and average thickness of aeolian sediment in this region is uncertain, but soil surveys suggest that some may be more than 2 m thick, although the thickness of much of it may not exceed 1 m. Furthermore, San Juan River is a probable contributing source, because the sand content decreases and silt content increases with increasing distance northeast.

c)   Columbia Plateau region

In this region, there are large tracts of loess in the Palouse area, then the Snake River Plain.

In Palouse region of eastern Washington and adjacent parts of Idaho and Oregon, loess may cover as much as 50,000 km^2. And it is as thick as 75 m. Such loess itself is thought to be derived primarily from fine-grained slack water sediments which are derived from cataclysmic floods of former proglacial Missoula Lake.

In the vicinity of the Snake River Plain in Idaho and western Wyoming, also adjacent uplands of Idaho, loess covers large areas to the north and south of the Snake River Plain, which is probably one of its major sources. Loess in this region has a thickness of up to 12 m in places, but most is 2 m or less.

d)   Alaska (USA) and Klondike, Yukon (Canada)

In the northwestern part of North America, loess is found in Alaska and the adjacent western parts of Yukon Territory of Canada.

Throughout Alaska, the thicknesses of the deposits range from a few centimeters in some areas to more than 60 m near Fairbanks. Loess deposits are thickest near rivers, with thicknesses decreasing rapidly with distance away from the rivers and downwind of valley dust sources. The transport and deposition of loess are processes that are still active today in Alaska, particularly along the Delta, Knik, Matanuska, and Copper Rivers, all of which drain mountain ranges with glaciers. Holocene loess is exposed in these valleys.

Other features can be described through 3 areas: Central Alaska, Itkillik (Northern Alaska), and Klondike (Yukon, Canada).

In central Alaska, Loess is widespread in the Fairbanks area, and much of the loess on north and northeast facing slopes has remained continuously frozen since it was deposited in the Pleistocene. The loess comprises both direct airfall silt on uplands and a combination of airfall and colluvially reworked loess in valleys. Sedimentologically, the central Alaskan loess tends to be massive, with stratification absent to indistinct, not only on uplands but in many lowlands.

The Itkillik (Northern Alaska) loess like the yedoma in northeastern Siberia, has been mapped as upland loess deposits, and occurs within 100–200 km of Holocene and modern loess deposits of the Prudhoe Bay region and the Colville River Delta.

Although loess deposits are widespread in western Yukon, they are thinner and less continuous than those in Alaska. Loess occurs in most valleys in southwestern and west-central Yukon, though the deposits have no distinctive surficial expression, and thus it is probably under represented relative to its true extent. The thickest and most extensive loess deposits in Yukon Territory occur in the unglaciated Klondike region of the Yukon River valley. These ice-rich loessal (or 'muck') deposits are also likened to Siberian yedoma.

4) South America

South American loess presents a broad geographic distribution extending across the Chaco–Pampean Plain of Argentina and neighboring areas of Uruguay, southern Brazil,

Paraguay, and the eastern Bolivia lowlands. Loess is extensive in the western Chaco and forms a wide belt in the eastern Pampas, grading into sand mantles and dune fields toward the west and southwest.

Thick loess deposits are located in the sub Andean mountainous area of northwestern Argentina (Tucumán); they also occur in the highland plains of Sierras Pampeanas (Pampas Mountains) of Córdoba and San Luis. In addition, loess has been reported in Tierra del Fuego and the eastern Andean piedmont, while Holocene peri desert loess has been reported in the Atacama Desert of Peru.

The thickest (~40–50 m) Quaternary loess and loess-like deposits are located in the northern Pampas with only the uppermost 10–15 m exposed (Muhs et al., 2014).

5) Australia and New Zealand

In Australia there is a scarcity of published work describing "loess" and a general view that soil derived from fine-grained aeolian sediment is quite restricted in distribution. Nowadays, a counterview has emerged that the hitherto described "parna" should be regarded as clayey, hot-climate loess.

At present on the Australian continent, there are two major wind paths emanating from the arid interior. One of these extends in an east–south easterly direction across the eastern states to the Tasman Sea, and the other extends in a northwesterly direction across Western Australia to the Indian Ocean. It is along the first so-called southeastern dust path that the majority of the identified Australian loess deposits exist. In particular, southern New South Wales and northern Victoria have prominent areas of clayey loess, and often calcareous soil derived from clayey loess (parna) deposits. In this area, the west part have widespread fine-grained loess (parna) deposits and the east part become patchy loess deposits.

In New Zealand, the climate is generally cooler and wetter, loess is widespread and very well described. Loess mantles are widely distributed across both the North and South Islands, about 10% of the country's land area (~26,000 km^2) is covered by loess at least a meter thick.

The range of loess mantle thickness is 0.5–6.0 m, but the thickest mantles are located in the southern parts of the North Island and the eastern and southern parts of the South Island.

On the North Island, loess deposits are most prominent in the Manawatu region (in Mid-south of North Island), where old terraces of cold-climate floodplains have been mantled by dust from nearby braided streambeds, and in the inland basins of Hawke's Bay (Southeastern North Island). On the South Island, loess deposits are most prominent on the high terraces of the Canterbury Plain (Eastern South Island), the North Canterbury and South Canterbury downlands, the Banks Peninsula (In Eastern Canterbury), and the South Otago downlands and Southland Plains (Southern South Island) (Muhs et al., 2014).

6) Africa and Arabian Peninsula

Loess in Africa and Arabian Peninsula is not as extensive and thick as in Europe, in North America and in China: it is patchy in nature. Yet loess is the parent material for some of the most fertile soils in these regions. Some characteristics are common to all reported desert loess sites in Africa and Arabian Peninsula: (1) loess bodies are located in semiarid to sub-humid climate regions; (2) at most sites, loess covers an area of 10^3–10^4 km^2 (Muhs et al., 2014; Crouvi et al., 2010).

The best-studied desert loess in northern Africa is located on the carbonate-rich Matmata Plateau in southern Tunisia. The Tunisian loess covers an estimated area of ~4,000 km^2 and reaches thicknesses of up to 20 m.

Loess deposits are also located on a carbonate-rich mountain range in northwestern Libya, between the Jefera Plain to the north and the Tripolitanian Plateau to the south. These deposits are considered to be a continuation of the Tunisian loess.

The best-known loess in the Sahel of Africa is the informally named Zaria loess, located on the Kano Plains in central-northern Nigeria. This is probably one of the largest desert loess regions known in the world, covering an area of 41,000 km^2 and located today in a tropical climate zone.

Scattered reports from other Sahelian countries suggest the presence of a noncontinuous loess belt, oriented west to east from southern Senegal through Guinea, Mali, Burkina Faso, Niger, and Nigeria to northern Cameroon; loess-like deposits are also found on the Canary Islands.

Loess in southern Africa is found mainly in northwestern Namibia, where it is located on a vast area between the Great Escarpment to the western coast.

The most prominent loess in the Arabian Peninsula is located in the Negev desert, southern Israel, covering an area of ~5,500 km^2. It mantles most of the exposed carbonate bedrock in the northern Negev and fills depressions and valleys farther south in the central Negev highlands.

Elsewhere in Arabian Peninsula, loess is found in northwestern Yemen, mainly located on the volcanic plateau of Sanaa-Dhamare-Taizz (1400–2700 m a.s.l). Loessial soils in Yemen are also reported from the arid Valley of Sadah, about 200 km to the north of the Sanaa plateau.

In the United Arab Emirates (UAE), desert loess reported near Ras Al Khaimah, as a ramp 3 km long against the Oman Mountains.

1.4.2 Distribution of loess in China

China loess covers a total area of ~635,280 km^2 (not including the Quaternary alluvial deposits with secondary loess in the North China Plain and Yangtze river basin) (Fig. 1.7), accounting for ~6.6% of China's land surface. Among them, the area of the primary loess is ~380,840 km^2, and ~254,440 km^2 of the secondary loess, respectively (Liu, 1965).

Loess in China is mainly distributed along the north of Kunlun Mountains, Qinling Mountains and Shandong Peninsula, and the south of Altai Mountains, Alxa, Erdos, Yinshan Mountains and Da Hinggan Mountains, forming a loess belt from west to east roughly. The eastern end of the loess belt is spread out in two directions to the north and south; the north part reaches the northern Songliao Plain of Northeast China, and the south part reaches the Shandong Peninsula and the middle and lower reaches of Yangtze River. In addition, Nanjing, Chengdu and the valley of Dadu River in Sichuan all have sporadic distributions of loess (Yang et al., 1988). Generally speaking, China loess belt mantles a latitude range from 30°N to 49°N and a longitude range from 75°E to 128°E. The area between latitude 34°N and 41°N of where the loess most developed,

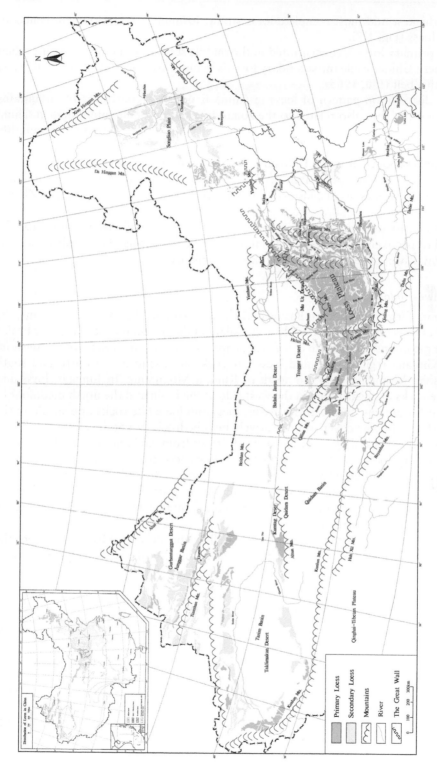

Figure 1.7 Distribution of loess in China (Modified from Li & Shi, 2017).

has the thickest and most complete loess strata, forming the development center of Chinese loess (Fig. 1.7).

The primary loess is concentrated in the middle section of the east-west loess belt in Northern China, while the secondary loess is mainly distributed in the east and west ends of this belt (Liu, 1965).

The distribution pattern of loess in China is not only controlled by mountains and topography, but also related to the zonal distribution of climate. Loess in China mainly lies in arid and semi-arid area of North China, where the lowest monthly temperature is less than 0°C, the annual mean precipitation between 250–500 mm, and annual evaporation above 1000 mm. The distribution of secondary loess can reach areas with annual mean precipitation of 750 mm. In the area with annual rainfall less than 250 mm, no loess exists, instead of deserts and Gobi (Liu, 1965; Wang & Zhang, 1980).

The distribution characteristics of loess in different regions of China can be depicted respectively from west to east: northwest inland basins, the middle reaches of the Yellow River, the eastern piedmont hills and plains.

1) Northwest inland basins

In Junggar, Tarim and Qaidam Basins in the west surrounded by steep mountains, the primary and secondary loess mainly accumulate in the mountains at margin of basins or in the piedmont plains, with consistent shapes of basins and an east-west extension.

In Xinjiang region, loess and loess like rocks are found in the north and south of Tianshan Mountains and the north of Kunlun Mountains. In Junggar, loess and loess like rocks are distributed in the west side of the basin and the north piedmont of Tianshan Mountains. In the Tarim Basin, loess and loess like rocks appear along the edge of the basin, and cover some relatively high bedrock mountains.

Hexi Corridor and Qaidam Basin are different from Xinjiang region; the primary and secondary loess are developed in the southeastern parts of both areas. This is related to the two different wind directions from Inner Mongolia.

In the Qaidam region, a small amount of the primary and secondary loess is distributed in Xiangride and the Burhan Budai mountains, both of which are located in the southeast side of the basin.

Hexi Corridor is located between the Qilian and Beishan Mountains; here the loess is gradually reduced westward after Wushaoling. But the secondary loess is very developed, and most are the front deposition of alluvial fans.

2) Middle reaches of the Yellow River

This area is roughly bordered by the Wushaoling Mountains to the west, the Taihang Mountains to the east, the Qinling Mountains to the south and the Great Wall to the north. It is geographically located in the middle reaches of Yellow River (from Longyangxia in Qinghai to Sanmenxia in Henan) and its tributaries. This area mantles between 34°N to 41°N in latitude and between 102°E and 114°E in longitude, with a north-south distance of ~700 km, and an east-west distance of ~1200 km.

In this region, the primary loess is distributed continuously in a large area, while the secondary loess is only found in valleys. The primary loess here is ~275,600 km², accounting for ~72.4% of the total area of the primary loess in China. Together with the secondary loess in this region, the total loess covered area is $\sim$278,000 km^2, $\sim$43.8% of the total loess area in China (Liu, 1965).

Except for Qinghai with little distribution of loess in this region, the loess is widely distributed in Gansu, Shaanxi, Shanxi and Henan, more developed than the Northeast, North and Northwest of China. Especially in the vast areas from the Liupan Mountains of Gansu to the Lvliang mountains of Shanxi, the loess distribution is wider and thicker than other places. In addition to a number of bedrock mountains higher than the loess depositional surface, the thick loess generally fills and covers the Tertiary and other old rock strata continuously, forming the platform (Yuan), ridge (Liang), hillock (Mao) and other loess landforms. Therefore, this region is named as the Loess Plateau, the home to the largest loess accumulation in the world and known to experience the most serious soil erosion.

The Loess Plateau is divided into three distinct sub-regions by a range of mountains aligned north to south: a) the western sub-region between the Wushaoling and the Liupan Mountains, where the loess is mainly distributed on mountain slopes, inter mountain basins and high terraces; the loess depositional surface basically reflects the undulating terrain of the basement; the underlying strata of loess are mainly composed of the Gansu Group in the Tertiary period; b) the central sub-region between the Liupan Mountains and the Lvliang Mountains, where the loess becomes a continuous layer covering the Jingle red clay in Pliocene and fills most of the former valleys and basins; the loess is more than 100 m thick and the sequence is complete; and c) the eastern sub-region between the Taihang Mountains and the Lvliang Mountains, where the loess covers the margin of the basins and the river terraces, and thin layer covers some ridges (Liu, 1985).

3) Eastern piedmont hills and plains

In Northeast China, the primary loess is only found in the southwest of Songliao Plain, which is connected with the desert in the north. The secondary loess is distributed in the contiguous area between the Songliao Plain and the mountainous area in the east of Northeast China.

In North China, the primary loess is mostly distributed on the piedmont low hills and high valley terraces, e.g., the south of Yanshan Mountains, the east of Taihang Mountains, the foothills of Taishan and Lushan Mountains, the north of Shandong peninsula and the basins, all having a wide distribution of loess. By comparison, the distribution of secondary loess is wider than the primary loess, common in lowlands along the front of loess covered mountains. Along the south of Yanshan Mountains via the Taihang Mountains and the east of Qinling Mountains, east to the mountains in mid-south Shandong and the north of Shandong peninsula, loess is exposed in such a vast area. Other places such as the north of Haihe River Plain and some area of Yellow River Plane also have the distribution of loess.

The primary and secondary loess in the North China Plain are more developed than the Northeast Plain, but inferior to the Loess Plateau.

Chinese loess is generally distributed between the altitude of 200 m and 2,400 m. In places such as the Kunlun mountains in the south of Tarim Basin, the distribution of loess can reach altitude 4,020 m. The loess with altitude more than 2000 m is mainly distributed in the west of Liupan Mountains in the middle reaches of Yellow River.

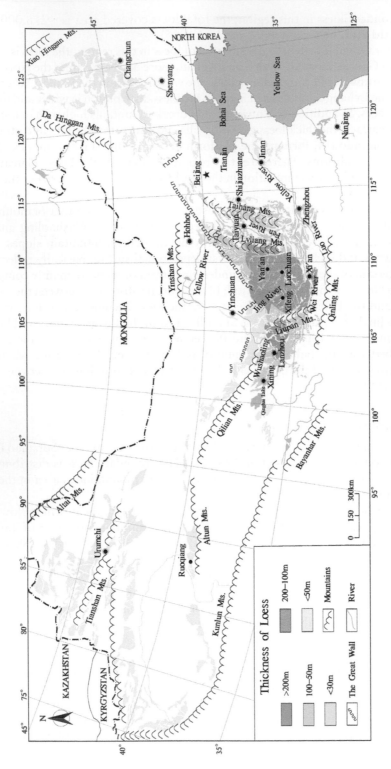

Figure 1.8 Thickness distribution of loess in China (Modified from Li & Shi, 2017).

Loess east of the Liupan Mountains on the Loess Plateau is located between altitude of 1000 m and 2000 m. Below 1000 m, loess is mainly distributed in some basins and plains in Eastern China, as well as in foothills in Western China. In general, the primary loess is located in high position, and the secondary loess in low position. The height of loess distribution in Xinjiang and the middle reaches of Yellow River are found to be related to the wind directions. In the windward slope, loess can be upraised to a higher position due to the stem of mountains. For example, the loess in the north slope of Liupan mountains reaches altitude 2070 m, and 1500 m in the south slope (Liu, 1965; Wang & Zhang, 1980; Li & Sun, 2005).

The thickness of loess deposits varies greatly in different regions across China (Fig. 1.8). The maximum thickness of loess is found in the middle reaches of Yellow River. In the west of Liupan Mountains, the north of Huajialing – Maxianshan Mountains to near Lanzhou, the thickness of loess ranges between 200 m and 300 m. From the east of Liupan Mountains to the west of Lvliang Mountains, the thickness of loess ranges between 100 m and 200 m. In the northern slope of Qilian Mountains, Kunlun Mountains and Tianshan Mountains, the thickness of loess is below 50 m. The thickness of loess in the Long'er Si Valley in the Tianshan Mountains is 120 m. The loess profile located in the south of Ganzi, Sichuan is also over 100 m thick. The loess in Westen Liaoning is about 100 m thick, and the loess in the Songliao plain is 40 m thick. Most of the loess in the North China Plain interacts with other alluvial strata, and the thickness is not large. According to current knowledge, the world's thickest loess is located in Jiuzhoutai, Lanzhou with a thickness of 335 m (Wang & Zhang, 1980; Li & Sun, 2005).

REFERENCES

Berg L.S. (1916) The origin of loess. *Izvestiya Russkogo Geograficheskogo Obshchestva*, 52, 579–646.

Busacca, A.J., Begét, J.E., Markewich, H.W., Muhs, D.R., Lancaster, N. & Sweeney, M.R. (2004) Eolian sediments. In: Gillespie, A.R., Porter, S.C. & Atwater, B.F. (eds.) *The Quaternary period in the United States*. Amsterdam, Elsevier. pp. 275–309.

Charpentier, J.D. (1841). Essai sur les glaciers. *Ducloux, Lausanne*.

Crouvi, O., Amit, R., Enzel, Y. & Gillespie, A.R. (2010) Active sand seas and the formation of desert loess. *Quaternary Science Review*, 29, 2087–2098.

Dodonov, A.E. (2007) Loess records | central Asia. In: Elias, S. & Mock, C. (eds.) *The encyclopedia of Quaternary sciences*. Amsterdam, Elsevier. pp. 1418–1429.

Eden, D.N. & Hammond, A.P. (2003) Dust accumulation in the New Zealand region since the last glacial maximum. *Quaternary Science Review*, 22, 2037–2052.

Frechen, M., Kehl, M., Rolf, C., Sarvati, R. & Skowronek, A. (2009) Loess chronology of the Caspian Lowland in Northern Iran. *Quaternary International*, 198(1–2), 220–233.

Haase, D., Fink, J., Haase, G., Ruske, R., Pécsi, M., Richter, H., Altermann, M. & Jäger, K.-D. (2007) Loess in Europe-its spatial distribution based on a European loess map, scale 1:2500000. *Quaternary Science Reviews*, 26(9–10), 1301–1312.

Hesse, P.P. & McTainsh, G.H. (2003) Australian dust deposits: modern processes and the Quaternary record. *Quaternary Science Review*, 22, 2007–2035.

Iii, E.A.B., Muhs, D.R., Roberts, H.M. & Wintle, A.G. (2003) Last glacial loess in the conterminous USA. *Quaternary Science Review*, 22(18), 1907–1946.

Jefferson, I.F., Evstatiev, D., Karastanev, D., Mavlyanova, N.G. & Smalley, I.J. (2003). Engineering geology of loess and loess-like deposits: a commentary on the russian literature. *Engineering Geology*, 68(3–4), 333–351.

Keilhack, K. (1920) Das Rätsel der Lössbildung. *Zeit. Deutsch. Geol. Gesell. B*, 72, 146–167.

Li, B.C. & Sun, J.Z. (2005) *Loess and environment in China* (In Chinese). Xi'an, Shaanxi Science and Technology Press.

Li, Y.R. & Shi, W.H. (2017) Review of provenance, transportation and distribution of loess. *Journal of Asian Earth Sciences* (under review).

Liu, T.S. (1964) *Loess on the Middle Reaches of the Yellow River* (In Chinese). Beijing, Science Press.

Liu, T.S. (1965) *The Deposition of Loess in China* (In Chinese). Beijing, Science Press.

Liu, T.S. (1985) *Loess and the Environment* (In Chinese). Beijing, China Ocean Press.

Lyell, C. (1834) Principles of Geology (III) (2nd ed). London: John Murray, Albemarle street.

Matsu'Ura, T., Miyagi, I. & Furusawa, A. (2011) Late quaternary cryptotephra detection and correlation in loess in northeastern Japan using cummingtonite geochemistry. *Quaternary Research*, 75(3), 624–635.

Muhs, D.R., Cattle, S.R., Crouvi, O., Rousseau, D.D., Sun, J.M. & Zárate, M.A. (2014) Loess Records. In: Knippertz, P. & Stuut, J.W. (eds.) *Mineral Dust: A key player in the Earth System*. Berlin, Springer Verlag, 411–441.

Murton, J.B., Goslar, T., Edwards, M.E., Bateman, M.D., Danilov, P.P., Savvinov, G.N., Gubin, S.V., Ghaleb, B., Haile, J., Kanevskiy, M., Lozhkin, A.V., Lupachev, A.V., Murton, D.K., Shur, Y., Tikhonov, A., Vasil'chuk, A.C., Vasil'chuk, Y.K. & Wolfe, S.A. (2015) Palaeo environmental interpretation of yedoma silt (ice complex) deposition as cold-climate loess, duvanny yar, northeast siberia. *Permafrost & Periglacial Processes*, 26(3), 208–288.

Nanjing Soil Research Institute, CAS. (1978) *Chinese soil* (In Chinese). Beijing, Science Press.

National Research Council (U.S.). Committee for the Study of Eolian Deposits (1952) *Pleistocene eolian deposits of the United States, Alaska, and parts of Canada .1:2500000*. Boulder, Colorado.

Обрлев, В.А. (1958) *Sand and loess issues* (In Chinese). Beijing, Science Press.

Péwé, T.L. (1975) *Quaternary Geology of Alaska*. Washington, DC, U.S. Government Printing Office.

Péwé, T.L. & Journaux, A. (1983) Origin and character of loesslike silt in unglaciated south-central yakutia, siberia, u.s.s.r. *Geological Survey Professional Paper*.

Pumpelly, R. (1867). *Geological Researches in China, Mongolia, and Japan: During the Years 1862-1865* (Vol. 15, No. 4). London, Smithsonian institution.

Richthofen, B.F. (1882) II.-On the mode of origin of the loess. *Geological Magazine*, 9(7), 293–305.

Rozycki. S.Z. (1991) *Loess and Loess-like Deposits*. Wroclaw, Ossolineum.

Smalley, I.J. (1995) Making the material: the formation of silt sized primary mineral particles for loess deposits. *Quaternary Science Reviews*, 14(7), 645–651.

Smalley, I.J. (1996) The properties of glacial loess and formation of loess deposits. *Journal of Sedimentary Research*, (3), 669–676.

Sun, J.M. (2002) Provenance of loess material and formation of loess deposits on the Chinese Loess Plateau. *Earth and Planetary Science Letters*. 203(3–4), 845–859.

Tanino, K., Hosono, M. & Watanabe, M. (2015) Distribution and formation of tephric-loess dunes in northern and eastern Japan. *Quaternary International*, 397, 234–249.

Virlet D'Aoust, P.T. (1857) Observations sur un terrain d'origine météorique ou de transport aérien qui existe au Mexique, et sur le phénomène des trombes de poussière auquel il doit principalement son origine. *Société géologique de France*, 2(2), 129–139.

Wang, Y.Y. & Zhang, Z.H. (1980) *Loess in Chinese* (In Chinese). Xi'an, Shaanxi people's Art Press.

Watanuki, T., Murray, A.S. & Tsukamoto, S. (2005) Quartz and polymineral luminescence dating of Japanese loess over the last 0.6 Ma: Comparison with an independent chronology. *Earth & Planetary Science Letters*, 240(3), 774–789.

Wright, J., Smith, B. & Whalley, B. (1998) Mechanisms of loess-sized quartz silt production and their relative effectiveness: laboratory simulations. *Geomorphology*, 23(1), 15–34.

Wright, J.S. (2001) 'Desert' loess versus 'glacial' loess quartz silt formation, source areas and sediment pathways in the formation of loess deposits. *Geomorphology*, 36, 231–256.

Yang, Q.Y., Zhang, B.P. & Zheng, D. (1988) On the boundary of the loess plateau (In Chinese). *Journal of Natural Resources*, 3(1), 9–15.

Zhu, X.M. (1983) Theory of the original soil formation (In Chinese). *Research of Soil and Water Conservation*, 2(4), 83–89.

Chapter 2

Loess landforms

2.1 INTRODUCTION

Landforms refer to various forms of the Earth's surface that result from internal and external geological processes over time. The Loess Plateau of China covers an area of 430,000 km^2 and is bounded by the Qilian Mountains to the west, the Taihang Mountains to the east, the Mu Us Desert to the north, and the Wei River Basin and Qinling Mountains to the south. After nearly 2.4 million years of continuous deposits (>100 m) and erosion, spectacular loess landforms have formed over this plateau (Zhang, 1983; Zhang, 1986). The loose nature of aeolian deposits has promoted the formation of landforms of various shapes in a relatively short time.

Loess landforms can be divided into three main types, namely, platform (including platform, residual platform and tableland), ridges, and hillocks. Figure 2.1 shows the regional distribution of these types of loess landforms across the Loess Plateau of China. The loess platforms are mainly distributed in Xifeng and Luochuan, which respectively belong to the Gansu and Shaanxi Provinces (Zhang, 1981; Zhang, 1986). The loess residual platforms are mainly located in the region bounded by the Baiyu Mountains to the south, the Mu Us Desert to the north, and the flanks of some loess platforms. The loess tablelands are mainly located on both sides of the Fen River and Wei River in the north of the Qinling Mountains. Loess ridges are mainly distributed in Huajialing between the Qilian Mountains and the Liupan Mountains, the peripheries of Xifeng platform, Luochuan platform and Ziwu Mountains, and the region between the Yellow River and the Lvliang Mountains. Loess hillocks are located in northern Lanzhou, Huanxian County; in the upstream of the Malian River; in Yan'an and Suide north of Shaanxi, and on both sides of the Yellow River at the boundaries between Shanxi and Shaanxi Provinces.

2.2 PLATFORMS

Loess platforms (Yuan in Chinese), which usually develop on basins, river terraces, or piedmont plains formed at the end of the Tertiary, are geomorphic units with a large area and a flatten-top after being eroded due to uplift (Zhang, 1983; Zhou et al., 2010). The area of a loess platform can reach up to several thousands of square kilometers. The slope of the top surface is generally 1–3 degrees, and increases gradually to the edge. Gullies with different sizes are distributed around loess platforms

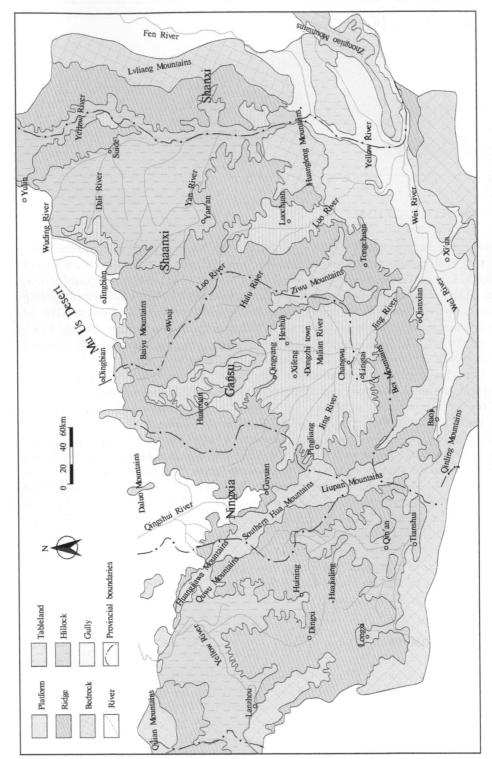

Figure 2.1 Types of loess landforms on the Loess Plateau of China (Modified from Zhang,1981).

Figure 2.2 Loess platform (Yuan) (Xifeng, Gansu Province).

(Xiao & Tang, 2007; Luo, 1956; Zhu, 2009) (Fig. 2.2). The largest gullies often run all the way from the top to the bottom of the platform, and dendritic drainage is formed by subsidiary branches.

The loess platforms developed on stepped river terraces are called loess tablelands (Tai-Yuan in Chinese). Tablelands are generally distributed on both sides of rivers in long strips, and their surfaces are inclined toward river valleys in a stepped shape (Fig. 2.3). Most of the lower part comprises alluvial and diluvial lacustrine deposits in the Tertiary or the Early Quaternary, and the upper part comprises aeolian loess and alluvial loess of the Quaternary. The tableland surface is complete and flat, with a slope generally below 5 degrees and slightly beyond 10 degrees. At the edges of loess tablelands, gullies are developed with a density of less than 1 km/km^2. All tablelands at different levels are connected by terrace scarps, and the relative elevation ranges from 50 m to 150 m. The edge of a tableland is mainly controlled by deep faults, which mainly drive the development of landslides and geological disasters. As the terraces were formed in different geological times, the overlying loess stratum is incomplete. Generally, early loess accumulation is obvious in the high-order terraces with an old age, and the stratum is relatively complete. The low-order terraces with a young age are only overlaid by new loess.

The gullies developed along the edges of loess platforms are cut by collapses, landslides, and water erosion. As the gullies become increasingly wide and long, the surface of the platform is dissected and developed into a residual loess platform (Can-Yuan in Chinese) (Zhang et al., 1987; Zhu, 2009) (Fig. 2.4). The slope of dissected surfaces can reach 3°–10°, with the density of gullies ranging from 2 km/km^2 to 3 km/km^2.

Figure 2.3 Loess tableland (Tai-Yuan) (Xi'an, Shaanxi).

Figure 2.4 Residual loess platform (Can-Yuan) (Xiangning, Shanxi).

2.3 RIDGES

Ridges (Liang in Chinese) are long strips of loess over low mountains and hilly terrains, and some are developed from platforms cut by gullies. According to the shape of their top surfaces, loess ridges can be divided into flatten-top ridges (Fig. 2.5) and inclined-top ridges (Fig. 2.6) (Wang & Zhang, 1980; Yang & Li, 2012). Flatten-top ridges

Figure 2.5   Flatten-top ridges (Liang) (Lvliang, Shanxi).

Figure 2.6   Inclined-top ridges (Liang) (Jixian, Shanxi).

generally have a length of several to tens of kilometers, a width of tens to hundreds of meters, and a slightly fluctuating elevation. The cross section (vertical to the divide) is slightly trapezoidal and turns into a ridge slope at the flank of ridges. The top surface of inclined-top ridges exhibits a small width, long slope, and a fish fin-like shape toward

Figure 2.7   Isolated hillock (Luochuan, Shaanxi).

both sides, where the side slope gradients range from 15 degrees to 35 degrees. The cross section shows a dome shape, and the divide shows obvious fluctuations.

2.4   HILLOCKS

Hillocks (Mao in Chinese) are formed and developed by the continued erosion of loess ridges or by deposition over existing dome-like terrains. These hillocks are oval or round in plain view. The angles of the upper slopes range from 3 to 10 degrees, which gradually increase to 15–35 degrees (Zhang, 1986; Zhang, 2000). A hillock may be isolated (Fig. 2.7) or may form an over lapping pattern with close proximity to other hillocks (Fig. 2.8). Isolated hillocks are rarely found on the edges of platforms and ridge zones. Concave sections forming saddles between hillocks are often the result of loess ridges cut by former gullies. The linearity of gully density in the area of hillocks can reach 5–8 km/km^2, which represents the area in the Loess Plateau with the most serious soil erosion.

Ridges and hillocks often roll together and form a gentle undulating loess landform referred to as loess hills.

2.5   GULLY

Gullies (Gou in Chinese) are formed by strong erosion and cutting of water at the edges of loess platforms, ridges, and hillocks (Fig. 2.9). In the same watershed, the gully density is greater and the gully banks are steeper on the sunny slope than those on the shady slope.

Figure 2.8 Continuous hillocks (Yan'an, Shaanxi).

Figure 2.9 Rills (Shouyang, Shanxi).

Figure 2.10 Shallow gullies (Xiangning, Shanxi).

After a storm, some scattered rills (Fig. 2.9) form in the upper part of platforms, ridges, and hillocks. These rills have a width of no more than 0.5 m, a depth of no more than 0.4 m, and a length ranging from a few centimeters to tens of meters. The slope of the bottom line of the longitudinal section is basically the same as the gradient of the slope. Rills will disappear after ploughing because of their small size. Shallow gullies (Fig. 2.10) with a cross section showing an inverted "Y" shape and a depth of 0.5–1.0 m are formed because of the collection of multiple streams and further downward cutting. The slope of the bottom line of the longitudinal section is slightly larger than the gradient of the slope. At the bottom of ridges and hillock slopes or in the vicinity of valley edges, cutter gullies form due to the collection, erosion, and cutting of overland flow. A cutter gully (Fig. 2.11) is generally narrow, deep, and steep, and it shows a stable position and a cross section with a sharp "V" shape. This type of gully has a depth of 10–50 m, width of up to tens of meters, and length of tens of meters to hundreds of meters.

With the further development of gullies with a cross section with a "V" shape, the cross section of gullies evolves into a "U" or box shape (Figs. 2.12 and 2.13). A gully shows a stable shape, a width of several hundred meters to several kilometers, and a depth of tens of meters to hundreds of meters. The bottom line of the longitudinal section is steep in the upper reaches and is gentle in the lower reaches. Gullies can be cut down into the loess layer of the Early and Middle Pleistocene or the laterite layer of the Pliocene, and a part of them cuts through the whole loess layer. The bottom of "U"-shaped gullies shows "V"-shaped gullies formed via modern underwater cutting. Seasonal or perennial water exists in gullies. Secondary loess slopes are usually developed in the gully slope, and secondary geomorphic units, such as sinkholes, loess walls, loess columns, and hanging gullies, are distributed along the sides of gullies.

Figure 2.11   Cutter gullies (Luochuan, Shaanxi).

Figure 2.12   Deep gullies (Luochuan, Shaanxi).

2.6 SECONDARY LANDFORMS

Loess landforms also include a variety of secondary landforms, such as loess dishes, loess craters (Fuller, 1922; Zhang, 1983; Zhang, 1986), loess walls, and loess columns,

Figure 2.13 Dry gully (Luochuan, Shaanxi).

which coexist with main landform types. They show distinct micro geomorphology features in the Loess Plateau.

2.6.1 Dish

Loess dishes refer to dishing hollows with a diameter of a few meters to tens of meters and a depth of usually only a few meters. They are formed by collapsibility (Yang & Li, 2012). They generally appear in gentle open loess platforms that are located tens of meters to one hundred meters away from loess gullies.

2.6.2 Crater

Subsurface erosion refers to the corrosion and erosion caused by the infiltration of surface runoff along the fissures and pores above the phreatic surface. Subsurface erosion action can destroy and change the original structure of loess and lead to the loss of particles and the formation of large pores and cavities. These conditions result in the collapse of loess strata and in the formation of various subsurface erosion landforms (Fuller, 1922; Zhu, 2009; Yang & Li, 2012). Subsurface erosion landforms mainly include loess craters, loess bridges, and hanging gullies.

Loess craters, also called loess sinkholes, are vertical caves formed by surface water flow and the subsequent subsurface erosion of loess joint fractures (Fig. 2.14). Loess sinkholes exhibit a circle-like bottom surface, a depth of several to tens of meters, a side wall of nearly 90 degrees, and the development of cluster shapes and strings of beads. Loess sinkholes are usually located in loess beams, loess platforms, residual loess platforms, and slope zones between gullies. Thus, sinkholes are a key to the

Figure 2.14 Loess sinkhole (Luochuan, Shaanxi).

retrogressive erosion of gullies. In addition, the bottom of a loess sinkhole is connected to another loess sinkhole and in direct contact with the slope and bottom of gullies. Loess sinkholes are the migration pathway of subsurface erosion water flow. Beaded sinkholes are the predecessor of some gullies. As a result of the existence of loess sinkholes, broken topography can be observed within 200 m along the lines of gullies.

When a relative impermeable layer exists, the downward seepage water produces lateral erosion, and the bottom of the loess sinkhole becomes connected to form a horizontal cave. If the top of the lateral cave is not collapsed, a loess bridge is formed. A loess bridge is a small landform unit with the shape of an arch bridge (Fig. 2.15). The span of a loess bridge can only reach about half of its height.

If the bottom of a beaded sinkhole in a cliff is connected and the top of the sinkhole is collapsed, then a semi-cylindrical vertical groove called a hanging gully (Fig. 2.16) is formed.

The development of loess cratering is closely related to vegetation, the depth of soil layer, and the grain size distribution of loess materials.

Figure 2.15 Loess bridges (Luochuan, Shaanxi).

The density of loess cratering is closely related to vegetation cover. The density of loess cratering is extremely large in areas with relatively sparse vegetation, and vice versa (Li et al., 2005; Li et al., 2009). Some loess sinkholes are covered with trees, and the lower limit of their formation age can be inferred from the age of tree growth.

If the depth of a loess layer is different, the characteristics, development density, and genesis of loess sinkholes also differ. As the number of loess sinkholes decreases, their scale becomes expansive, and their characteristics become obvious. A surface sinkhole is biogenic, a shallow sinkhole is related to the vertical differentiation of the anti-erosion ability of the loess stratum and various joints and fissures, and a middle sinkhole is mainly related to the joint fissure development in loess. The direction of the long axis of deep and extremely deep sinkholes is basically the same as the dominant direction of the structural joints in loess (Wang & Wang, 1993; Peng et al., 2007; Li et al., 2009).

In the Loess Plateau of China, the development degree of loess sinkholes is closely related to the grain size distribution of loess materials, and grain size distribution is the main cause of the zonal distribution of loess sinkholes. Loess sinkholes are mainly developed in the silty loess zone roughly bounded by the two black lines in Figure 2.17, and are rarely developed in the sandy loess zone to the north and clayey loess zone to the south. This feature is due to the high sand content of sand loess. The surface of sand loess is too loose and prone to disintegration and collapse; thus, loess sinkholes with overhead structures do not easily form in space. The clayey loess layer exhibits a high clay content, low collapsibility, low disintegration, and low permeability, all

Figure 2.16 Hanging gully (Luochuan, Shaanxi).

of which increase the erosion resistance of the layer. Therefore, loess does not easily erode, and the number of loess sinkholes in the clayey loess zone is low (Li et al., 2005; Li et al., 2010; Yang & Li, 2012).

2.6.3 Wall and column

Loess walls (Qiang in Chinese) or loess columns (Zhu in Chinese) (Figs. 2.18 and 2.19) are residual loess after the erosion action of water flow along the loess vertical joint and collapse. They are usually distributed at the sides of loess platforms, loess ridges, and protruding parts of gullies. The loess wall between two small loess gullies is formed because of the head erosion and lateral erosion of both sides of a loess gully

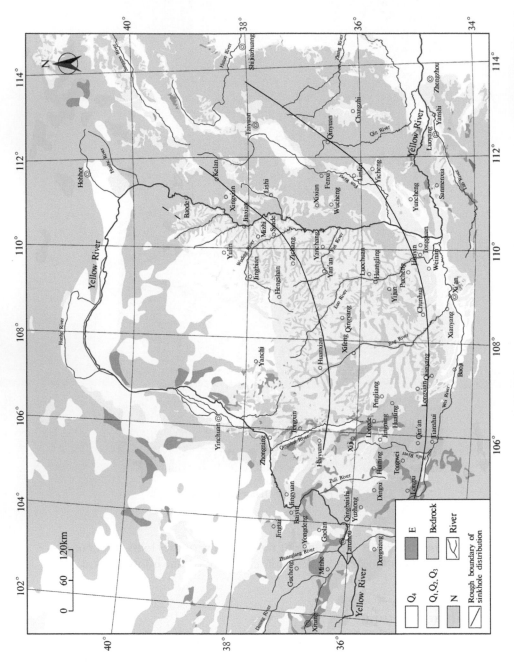

Figure 2.17 Zoning map of loess grain sizes and loess sinkhole (Modified from Peng et al., 2007). Two black lines indicate boundaries of different loess zones.

Figure 2.18 Loess wall (Qiang) (Luochuan, Shaanxi).

and the collapse of wall soil. In addition, loess walls or loess columns are formed via the erosion of soil by water flow along the edges of gully landforms or the joints parallel to gully walls. The length and height of a loess wall is generally a few meters to tens of meters, and the width is a few tens of centimeters to a few meters. In addition, loess walls with vertical joints or cracks can be transformed into loess columns under the conditions of landslide and collapse. Loess columns can exhibit tower, spire, and cone shapes. The surface of loess columns flakes, making them increasingly short and thin until they finally disappear. A plurality of loess columns clustered in the same place forms a loess forest.

2.7 FORMATION AND EVOLUTION OF LOESS LANDFORM

The present landform of the Loess Plateau of China was formed and developed gradually over a long geological time. The prototype was formed at the end of the Yanshan movement between the early Tertiary period and the end of the Cretaceous period about 70 million years ago. During this period, the Huanglong Mountain and Ziwu

Figure 2.19 Loess column (Zhu) (Left: Luochuan, Shaanxi; Right: Yuci, Shanxi).

Mountain exposed their surfaces, and the Wei River appeared because of the large-scale uplifting of the Erdos syncline platform. In the Miocene epoch of the Neogene, the Ordos platform rose rapidly into the plateau, with the terrain tilting from northwest to southeast. The Qinling Mountains had a sharp uplift, and the Fen-wei graben deepened. At the end of the Pliocene and early Pleistocene, the Wei River graben showed a stepped collapse, and a series of terraces formed as a result (Huang, 1945). At that time, the ancient topography of the loess landform was formed. On this basis of this ancient topography, the landform today was formed under the common effects of the accumulation and erosion of loess, differences in neotectonic movements since the Quaternary, human activities, and other external forces, such as water flow, frost weathering, rainfall, and gravity.

The development of loess landforms is accompanied by the accumulation and erosion of loess, which result in the multi-period change of loess landforms. On the large scale, loess covers the underlying topography and possesses the basic morphological characteristics of ancient landforms, but the relative height of the relief now is smaller than that of the underlying bedrocks. Obviously, the inheritance of loess landforms is only a type of general and contoured macroscopic characteristic. In fact, the accumulation and erosion occurring on a variety of ancient landforms prompt a certain degree of transformation among ancient landforms, and such transformation occurs between different types of landforms.

The edge of a loess platform on a plan resembles a flower because of the dense distribution of gullies on the side of the loess platform, the rapid head erosion of a gully, and the development of a gully from the surroundings to the loess platform. The edge zone of the platform gradually disintegrates, and a broken platform is formed as a result of the constant erosion of gullies to the platform surface. A broken platform is the transitional form between a loess platform and loess hillock. A broken platform can be further divided into loess hills. The loess hill at the edge of the platform is the product of the evolution of the loess platform.

Loess hills comprise loess ridges and hillocks with different scales. The loess ridges and hillocks are related in terms of distribution. Generally, loess ridges and hillocks are formed in two modes. First, loess ridges and hillocks develop depending on the underlying ancient hills. Second, they are the residual products of the erosion of the side of a loess platform. The general trend of evolution of loess hills is as follows: as the loess ridges and hillocks further break, the surface gradually flattens.

The late evolution of loess landforms resulted from the continuous disintegration of loess platforms, ridges, and hillocks. A loess platform may evolve from platforms to broken platforms to loess hills. In addition, a secondary loess landform also exhibits the same characteristics. For example, loess bridges and hanging gullies are the result of the erosion of loess sinkholes.

REFERENCES

Fuller, M.L. (1922) Some unusual erosion features in the loess of China. *Geographical Review*, 12(4), 570–583.

Huang, T.K. (1945) *On Major Tectonic Forms of China*. Geological Memoirs, Series A, 20. Nanjing, Central Geological Survey.

Li, X.A., Peng, J.B., Chen, Z.X., Kang, J.H. & Li, L. (2009) Erosion physiognomy of loess cave underground and comprehensive analysis on its advantages and disadvantages (In Chinese). *Journal of Xi'an University of Science and Technology*, 29(1), 78–83.

Li, X.A., Peng, J.B., Zheng, S.Y., Chen, Z.X. & Tian, A.J. (2005) A study on origin of loess caves in loessal plateau (In Chinese). *Highway*, (11), 142–146.

Li, X.A., Song, Y.X. & Ye, W.J. (2010) *Erosion of Engineering Geology of Loess Caves* (In Chinese). Shanghai, Tongji University Press.

Luo, L.X. (1956) A tentative classification of landforms in the loess plateau. *Geography Journal*, 22(3), 201–222.

Peng, J.B., Li, X.A., Fan, W., Chen, Z.X., Su, S.R., Song, Y.H., Lu, Q.Z., Deng, Y.H., Chen, L.W. & Sun, P. (2007) Classification and development regularity of the loess cave in Loess Plateau region (In Chinese). *Earth Science Frontiers*, 14(6), 234–244.

Wang, J.M. & Wang, J. (1993) Loess erosion landform and loess tectonic joints in the Middle and Southern Hebei (In Chinese), *Geographical Research*, 13(1).

Wang, Y.Y. & Zhang, Z.H. (1980) *Chinese Loess* (In Chinese). Xi'an, Shaanxi people's Fine Arts Publishing House.

Xiao, C.C. & Tang, G.A. (2007) The classification of shoulder line of valley in the loess landform (In Chinese). *Arid Land Geography*, 30(5), 646–653.

Yang, J.C. & Li, Y.L. (2012) *Principles of Geomorphology* (In Chinese). The third edition. Beijing, Peking University Press.

Zhang, Z.H. (1981) Regional geological features and modern erosion of the Loess Plateau in China (In Chinese). *Acta Geologica Sinica*, (4), 308–311.

Zhang, Z.H. (1983) The compilation principle of landscape type map of Chinese Loess Plateau (In Chinese). *Hydrogeology & Engineering Geology*, (2), 29–33.

Zhang, Z.H., Institute of hydrogeology and engineering geology of Chinese Academy of Geological Sciences. (1986) *Landscape type Map and Instructions of Chinese Loess Plateau* (In Chinese), 1:500000. Beijing, Geological Press.

Zhang, Z.H. (2000) *Nine Greet Bends with Miles of Sands in Yellow River – The Yellow River and the Loess Plateau* (In Chinese). Beijing, Tsinghua University Press and Jinan University Press.

Zhang, Z.H., Zhang, Z.Y. & Wang, Y.S. (1987) The basic geological problems on China loess (In Chinese). *Acta Geologica Sinica*, (4), 362–374.

Zhou, Y., Tang, G.A., Yang, X., Xiao, C.C., Zhang, Y. & Luo, M.L. (2010) Positive and negative terrains on northern Shaanxi Loess Plateau. *Journal of Geographical Sciences*, 20(1), 64–76.

Zhu, S.G. (2009) *The West Landmarks: Loess Plateau* (In Chinese). Shanghai, Shanghai Scientific and Technical Literature Publishing House.

Chapter 3

Loess microstructure

In previous studies of loess, macroscopic structure generally refers to the structure observed with unaided eyes, whereas microstructure refers to the structure observed under a microscope. At the macroscale, loess is a deposit of yellow-tawny to brownish loose silty sediment. It has no bedding structure. At the microscale, loess is composed of individual mineral grains, mineral aggregates, and pores.

3.1 MICROSTRUCTURAL CHARACTERISTICS

Particle size and shape, particle contacts, role of cementation, and pores, are the focal points, and an optical microscope and/or scanning electron microscope (SEM) are used in previous studies of the loess microstructure.

3.1.1 Particle size and shape

Individual particles are generally characterized by size and shape; particle size distribution is represented by a cumulative distribution curve, whereas the shape of particles can be characterized by the morphology of the projected image.

Granulometric composition

The particle size composition of loess shows no strong variation across time and region. A majority of loess particles are less than 0.25 mm in diameter and loess generally has three particle fractions: sand (>0.05 mm), silt (0.05–0.005 mm), and clay (<0.005 mm). Figure 3.1 shows the particle size composition of the Malan loess in different regions. It is seen that content ranges of sand, silt, and clay are 10%–50%, 40%–70%, and less than 40%, respectively.

Particle shape

Abrasion during the transport process changes the shape of loess particles. The abrasion degree of particles depends on the type of transportation medium as well as the intensity and duration of transportation. The abrasion degree can be described by the angularity index, which describes the sharpness degree of the edges of a particle. Loess

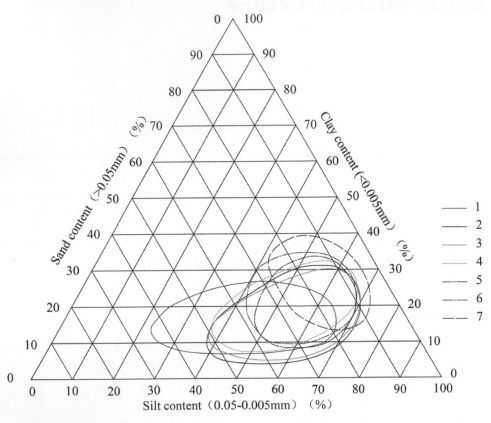

Figure 3.1 Comparison of particle size compositions of the Malan loess in different regions: 1. Qaidam (Qinghai); 2. Longdong (Gansu); 3. Shaanxi; 4. Longzhong (Gansu); 5. Shanxi; 6. Longxi (Gansu) and Qinghai eastern; and 7. Shandong. (Redrawn from Liu, 1965).

particles can be divided into four classes according to their angularity (Zhang, 1964), as shown in Table 3.1. For example, angular particles have straight outlines, sharp edges, and rough surfaces.

3.1.2   Occurrences of particles

Loess particles can be categorized in terms of their occurrence, including individual grain, coated grain, aggregate, and aggregate cluster (Fig. 3.2). Sand (>0.05 mm in diameter) and coarse silt (>0.01 mm in diameter) normally occur as individual grains (Figs. 3.2a & 3.3a) (Liu & Zhang, 1962; Liu, 1966; Gao, 1980a; Xing, 2004; Wang, 2010). Coated grains have surfaces covered with amorphous material (Fig. 3.2b). Aggregate (Figs. 3.2c & 3.3b) is a collection of clay and/or fine silt particles with diameters ranging from 2 μm to 50 μm held together by cementing materials (Liu, 1965; Feng & Zheng, 1982; Gao, 1980a, 1980b; Li, 1997; Xing, 2004). Aggregate

Table 3.1 Particle shape features.

Angularity	Outline	Surface roughness	Sketch (Zhang, 1964)
Angular	straight	rough	
Subangular	straight and smooth	rough dominate	
Subrounded	smooth and straight	smooth dominate	
Rounded	smooth	smooth	

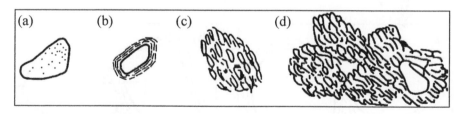

Figure 3.2 Loess particle occurrences: (a) individual grain; (b) coated grain; (c) aggregate; and (d) aggregate cluster (Gao, 1980a).

cluster (Figs. 3.2d & 3.3c) is a collection of individual grains and aggregates held together by cementing materials. Soluble salts, organic matter, free oxides, carbonates, and clay minerals are the main cementing agents. The diameters of loess aggregates generally range from 30 μm to 70 μm, whereas those of clusters range from 50 μm to 100 μm. The internal structure of aggregates and aggregate clusters are closely related to the degree of the leaching of calcium carbonate. The aggregates cemented by microcrystalline calcium carbonate are firm and rigid, whereas those cemented by clay and other materials are soft.

Figure 3.3 SEM images of (a) individual grain; (b) aggregate; and (c) aggregate cluster (Modified from Wang, 2010).

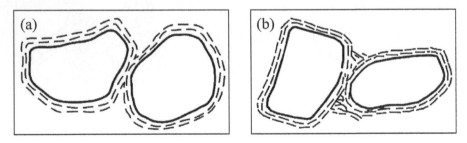

Figure 3.4 Particle contact relationship: (a) point contact; and (b) surface cementation (Gao, 1980a).

3.1.3 Particle contacts

Two forms of particle contacts exist in loess, namely, point contact and surface cementation. In point contact (Fig. 3.4a), particles directly contact each other, but the contact surface between particles is very small. Contact generally occurs between rigid aggregates or individual grains. Only a few salt crystals, clay, or other cementing materials

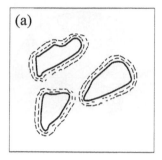
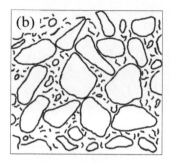

Figure 3.5 Forms of cementation in loess: (a) thin film; (b) interstitial; and (c) matrix (Lei, 1988).

reinforce the contact point, although aggregates and individual grains are normally wrapped by a layer of clay film, salt crystal film, or weathered crust. In surface cementation (Fig. 3.4b), the contact area between the particles is larger than that in point contact, and the cementation materials are thicker (Gao, 1980a).

3.1.4 Cementation characteristics

Clay minerals, soluble salts, organic matter, free oxides, and carbonates are the main cementing agents in loess (Liu & Zhang, 1962). Illite, montmorillonite, and kaolinite are the main clay minerals in loess. Clay minerals adhere to the surfaces of loess particles and the contacts between particles. Being hydrophilic, clay minerals exhibit swell–shrink behavior, and their bonding strength often change with water content. Therefore, cementation is strong under dry conditions but weak when saturated. Soluble salts can promote the aggregation of clay minerals, but these salts are less stable and easily leached out. Organic matter in loess occurs in minute quantities and exhibits high dispersion as well as strong hydrophilic and adsorption properties. Free oxides occur in two states, namely, microcrystalline and amorphous. In the microcrystalline state, free oxides do not contribute to cementation but function as small grains, but in the amorphous state, free oxides adhere to particle surfaces forming cementation (Zhang & Qu 2005; Wang et al., 1992). Calcium carbonates are classified as primary calcium carbonates or secondary calcium carbonates according to their formation (Gao, 1980a; Lei, 1983; Cai, 2014). Primary calcium carbonates mainly occur as individual particles and do not contribute to cementation. Secondary calcium carbonates are normally formed by weathering and leaching, and they can function as a cementing agent when they accumulate between the contacts and the particle surfaces.

Cementation in loess occurs in three forms, namely, thin film, interstitial filling, and matrix (Fig. 3.5) (Wang & Teng, 1982; Lei, 1983; Zhu, 1965). Cementing materials, which adhere to the surfaces of coarse grains, form a thin film (Fig. 3.5a). Interstitial filling is distributed in irregular pores formed by particles with contacts (Fig. 3.5b). When coarse particles float with little or no contacts within a base of cement, matrix cementation is formed (Fig. 3.5c).

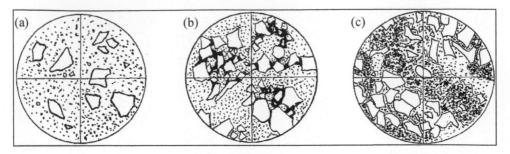

Figure 3.6 Coarse particle arrangements: (a) dispersed distribution; (b) contact distribution; and (c) porphyritic distribution (Zhang, 1964).

3.1.5 Particle arrangements and pore characteristics

The engineering properties of loess, e.g. collapsibility, compressibility, and permeability, are mainly determined by its structure. The internal structure of loess can be characterized by the spatial distribution and arrangement of particles and pores.

The coarse particles in loess are arranged or distributed in three different configurations (Fig. 3.6):

a) *Dispersed distribution*: The coarse particles form a fraction of the whole, and nearly no contacts exist between the coarse particles (Fig. 3.6a).
b) *Contact distribution*: The coarse particles are generally in point contact within a matrix of fine particles (Fig. 3.6b).
c) *Porphyritic distribution*: The coarse particles are unevenly distributed, and they form groups. Within groups, these particles are in contact with each other (Fig. 3.6c).

Pores in loess have two origins: primary and secondary (Table 3.2). Primary pores are formed in the process of accumulation. Examples of primary pores are intergranular pores, which include trellis pores and mosaic pores, and cement pores. Intergranular pores constitute 45%–85% of a body of pores. Cement pores usually account for 5%–35%. The loading conditions during and after burial and the climatic conditions directly affect the percentage of intergranular pores. When the loading of loess is high, the soil is dense, and intergranular pores are fewer. Intergranular pores can be easily formed and preserved under dry conditions. Secondary pores are formed by the actions of biological, biochemical, physical, and chemical processes during the pedogenesis of loess (Lei, 1983, 1988). Secondary pores normally constitute less than 50%, depending on the buried depth. In the vertical sequence of loess deposit, the degree of compaction increases, and the secondary pores (e.g., root channels and worm channels) significantly decrease with an increase in geostatic stress.

Biological pores, cracks, dissolution pores, and inter-aggregate pores are usually large, that is, with diameters larger than 0.016 mm. Trellis pores and intercrystalline pores are usually medium in diameter (0.004–0.016 mm). Mosaic pores are small (0.001–0.004 mm in diameter), and cement pores are microscopic (<0.001 mm in diameter).

Table 3.2 Types and characteristics of pores.

Types			Definition	Simple view	Characteristics
Primary	Intergranular pores	Trellis pores	Pores formed by the loose accumulation of coarse particles.	(Lei, 1987)	Pore diameters are generally larger than those of the surrounding particles. Minimal to no cementation occurs around the walls of the pores; thus, the stability of trellis pores is poor. Trellis pores are varied and irregular in shape. Their percentage significantly decreases from northwest to southeast across the Loess Plateau. They account for more than 50% of pores in sandy loess but less than 24% in clayey loess.
		Mosaic pores	Pores among well-packed coarse particles.	(Lei, 1987)	Mosaic pores are in the form of irregular tubes whose diameters are usually smaller than the diameters of the surrounding particles. The stability and connectivity of mosaic pores are better than those of trellis pores. The fraction of mosaic pores decreases from northwest to southeast across the Loess Plateau. They account for less than 30% of pores in sandy loess and less than 23% in clayey loess.
Primary	Cement pores		Pores distributed in cement.	(Wang, 2010)	Cement pores are abundant, but they have very small diameters (<0.002 mm) and various shapes. These pores are enclosed and randomly distributed. Their content is proportional to that of clay. They account for approximately 35% of pores in clayey loess and less than 5% in sandy loess.

(continued)

Table 3.2 Continued.

Types		Definition	Simple view	Characteristics
Secondary	Biological pores	Root channels	Residual pores after the disappearance of plant roots.	Root channels are nearly circular in cross section. Root channels are usually vertical in the section. Hole walls are usually lined by secondary calcium carbonate and other materials, which form a hard shell. Root channels account for less than 5% of pores in sandy loess and approximately 20% in clayey loess.
		Worm channels	Wormholes	Worm channels are curved and nearly circular in cross section. The walls of worm channels are softer than those of root channels; thus, these channels are easily destroyed. The diameters of worm channels range from 0.5 mm to 5 mm. Worm channels are more common in the top layers of the Malan loess. Worm channels in clayey loess are more than those in silty loess and sandy loess.
Secondary	Biological pores	Burrows	Large pipes created by burrowing animals.	Animal burrows are cylindrical. They are usually larger than 5 cm in diameter. Their walls, which are unstable, easily collapse. They are common in the top layers of the Malan loess.
		Cracks	Fractures formed by shrinkage upon drying.	Cracks are irregular and spindle shaped. Often, cracks extend as curved or broken line segments. Sometimes secondary clay minerals are deposited along the edges. Cracks are rare in sandy loess but common in clayey loess.

(Wang, 2010)

(Wang, 2010)

	Dissolution cavities	Residual cavities after dissolution of soluble salt minerals.		Dissolution cavities are radial, enclosed, and large. Clay materials coat the cavity walls and exist in the pore spaces. Dissolution cavities are mainly distributed in Late Pleistocene clayey loess and silty loess.
Secondary	Inter-aggregate pores	Pores between aggregates or aggregate clusters.	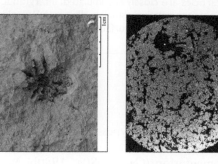 (Wang, 2010)	Inter-aggregate pores are found in thin sections, and they form an irregular web. They are more common in clayey loess.
	Intercrystalline pores	Pores formed by the recrystallization of soluble salt minerals, such as gypsum and calcite.	(Gao, 1980a)	The diameters of intercrystalline pores range from 0.005 mm to 0.01 mm. These pores are irregular in shape and enclosed. They are rare and found mainly in clayey loess.

Table 3.3 Classification of loess microstructure.

Main types	Subgroups	Characteristics
Poorly cemented	Granular, trellis, contact	Particles are clean and loosely accumulated with a small contact area and occasional cementation.
	Granular, trellis–mosaic, contact	Particles are clean and closely packed, inducing mainly surface contact and occasional cementa-
	Granular, mosaic, contact	tion. Intergranular pores are small in diameter.
Semi-cemented	Granular, mosaic, contact–cement	Particles are loosely accumulated. Particles and pores are not all clean, with cement materials on
	Granular, trellis, cement	their surfaces.
	Granular, trellis–mosaic, contact-cement	Particles are closely packed. Both pores and particles are not very clean, with cement
	Granular, trellis–mosaic, cement	materials on their surfaces.
	Granular, mosaic, cement	
Cemented	Granular–aggregate, trellis, cement	Particles are loosely accumulated. Intergranular pores are relatively large.
	Granular–aggregate, trellis–mosaic, cement	Particles are not clean as cement materials adhere to surfaces.
	Granular–aggregate, mosaic, cement	Particles are closely accumulated. Intergranular pores are relatively small in diameter. Particles
	Aggregate, mosaic, cement	are not clean, with surfaces covered with cement materials.

3.2 CLASSIFICATION OF MICROSTRUCTURE

The microstructure of loess can be divided into three main types under which several structural subgroups exist (Table 3.3), depending on the shape and arrangement of particles, form of contact, and degree of cementation (Gao, 1980a, 1980b; Wang, 1982; Lei, 1983, 1987, 1988, 1989). In Table 3.3, granular, trellis, contact loess particles are mainly individual grains, intergranular pores are mainly trellis pores, and particles are mainly in a point contact.

3.3 PARTICLE SIZE ZONING ACROSS THE LOESS PLATEAU

Particle and pore size compositions of loess in China show a regular distribution pattern with distance from the source area. The Loess Plateau is divided according to the loess grain size composition into three zones, namely, sandy loess zone, silty loess zone, and clayey loess zone (Liu, 1962, 1965, 1966; Zhu, 1963, 1964) (Fig. 3.7).

I. Sandy loess zone: It extends from the south of the Mu Us Desert to the north of Tongxin (Ningxia), Jiaxian (Shaanxi), and Xingxian (Shanxi) Counties (Fig. 3.7). In this zone, the mean diameter (Md) of loess particles ranges from 0.026 mm to 0.076 mm. The content of sand is generally high, with a wide range varying from 23.6% to 72.4%, rarely falling below 35%, and 41.25% on average. By contrast, the clay content in this zone ranges from 7.0% to 20.7%, with 16.81% on average.

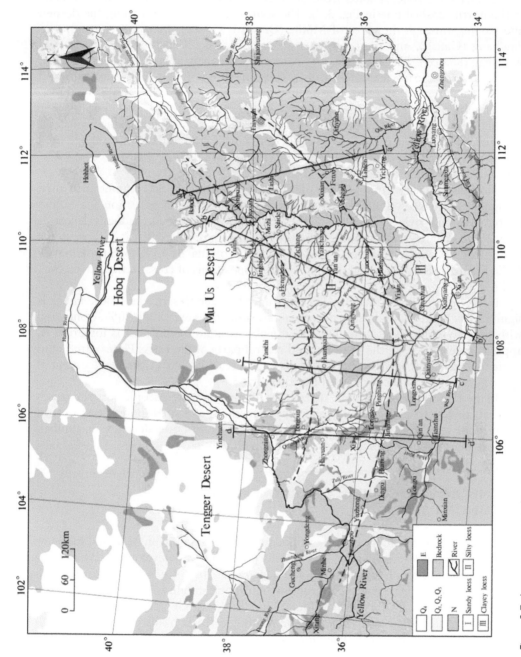

Figure 3.7 Loess zones according to particle size distributions; boundaries are indicated by dashed lines (Modified from Liu, 1985).

The silt content in the sandy loess zone is lower than those in other zones, being 41.25% on average.

II. Silty loess zone: It starts from the southern boundary of the sandy loess zone in the north (dashed line in Fig. 3.7). The southern boundary lies along the dashed line passing through the cities of Minhe (Qinghai) County, Lanzhou (Gansu), and Pingliang (Gansu), and the counties of Yijun (Shaanxi), Luochuan (Shaanxi), and Wucheng (Shanxi). The silty loess zone constitutes the main area of the Loess Plateau. The Md of loess particles ranges from 0.016 mm to 0.032 mm. The sand, silt, and clay content ranges in this zone are 11.1%–41.4%, 46.8%–70.0%, and 7%–30.4%, respectively.

III. Clayey loess zone: The southern boundary of the clayey loess zone is not well defined (Fig. 3.7). The particle Md in this zone ranges from 0.018 mm to 0.027 mm. The sand, silt, and clay contents ranges in this zone are 11.1%–29.1%, 45.5%–68.9%, and 17.9%–35.75%, respectively.

The granulometric compositions of the Malan loess in different regions are shown in Figure 3.8. As shown in Figure 3.8a, which represents the cross section a–a′ across western Shanxi Province in Figure 3.7, the sand fraction of the loess gradually decreases from approximately 60% at Bode in the north to approximately 13% at Yicheng in the south, whereas the clay fraction increases from approximately 15% to approximately 34%. Figures 3.8b–d represent the grain size compositions of cross sections b–b′, c–c′, and d–d′.

In the area close to the Mu Us Desert and Tengger Desert, the sand content is significantly higher than those in other regions, sometimes exceeding 70%. Compared with the clay contents in other regions, the clay content in this area is low, that is, approximately 10%. In the southern and central parts of the Loess Plateau, the sand content is generally less than 30%, occasionally reaching 40%; the clay content is approximately 20%, locally exceeding 30% (Fig. 3.8).

In the Loess Plateau of China, the volume of large and microscopic pores increases from northwest to southeast, whereas that of medium and small pores decreases. In the vertical sequence of loess deposit, large and medium pores decrease in number, whereas microscopic pores gradually increase with depth.

3.4 MICROSTRUCTURAL CHARACTERISTICS UNDER SEM

The microstructure of loess exerts a significant influence on the physical and mechanical properties of loess. In sandy loess, trellis pores are well developed, clay content is low, and cementation is poor. With these characteristics, sandy loess presents high collapse potential, high water permeability, and poor stability. The area covered by sandy loess is a zone of extensive slumps. In silty loess, clay content is higher and cementation degree is better than those in sandy loess, but trellis pores are still well developed to a certain extent. Therefore, sandy loess demonstrates good permeability and collapsibility and less frequent slumps. In clayey loess, clay content is higher and interparticle cohesion is stronger than those in sandy loess and silty loess. Clayey loess is well cemented, and it exhibits poor water permeability and collapsibility. This zone has only a few slumps.

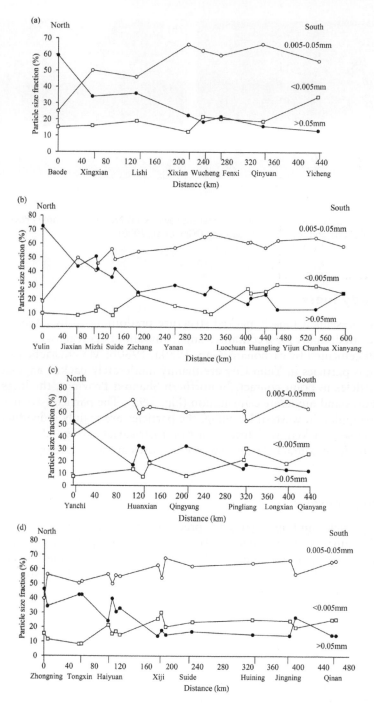

Figure 3.8 Variations in the grain size composition of the Malan loess along the four transects across the three loess zones: (a) a–a′ cross section; (b) b–b′ cross section; (c) c–c′ cross section; and (d) d–d′ cross section (see in Fig. 3.7) (Liu, 1965).

Figure 3.9 SEM images of the Malan loess in different regions: (a) Northern Shaanxi Province (Zhang & Qu, 2005); and (b) Western Liaoning (Xing et al., 2008).

3.4.1 Sandy loess

Sandy loess presents a loosely packed structure predominated by coarse silt, whose grains are coated; sandy loess contains a small amount of aggregates composed of clay and fine silt cemented by carbonate (Gao, 1980a, 1980b; МУстафасВ, 1984).

The loess particles in Yulin City are mainly moderately packed angular and sub-angular particles in point contact. In northern Shaanxi Province, the loess structure has large pores and moderate cementation (Fig. 3.9a). The particles are mainly angular and subangular. In western Liaoning, the particles are mainly individual particles, which are loosely packed, with little debris on their surfaces (Fig. 3.9b). The particles are mainly in point contact. Trellis and mosaic pores are well developed.

3.4.2 Silty loess

Individual particles and aggregates are predominant in silty loess. The particles are mainly subangular and subrounded. Trellis and mosaic pores are well developed. Cementation is mostly intergranular. Table 3.4 shows the specific microstructural characteristics under SEM in 17 different regions in the provinces of Gansu, Shaanxi, and Shanxi (Fig. 3.10).

3.4.3 Clayey loess

The particles of clayey loess are mainly in aggregate form and moderately to well packed and cemented. Table 3.5 shows the specific microstructural characteristics under SEM in 15 different regions in the provinces of Gansu, Shaanxi, Shanxi, and Henan (Fig. 3.11).

Table 3.4 SEM structures of the Malan loess in certain areas of the silty loess zone.

Area	Characteristics	Sample view and source
Xining	The particles are mostly individual particles. The cementation is moderate, producing some aggregates. The clay adheres to particle surfaces. The sizes of trellis pores are similar to those of the surrounding particles.	 ×500 100 µm (Deng, 2009)
Yongdeng	The particles are mainly subangular loosely packed individual particles with numerous large-sized trellis pores. The salt cementation is moderate.	 100 µm (Li et al., 2005)
Lanzhou	The particles are mostly subangular and moderately packed individual particles. A few particles contact with moderate cementation. Trellis and mosaic pores exist.	 ×300 200 µm (Wang, 2010)

(continued)

Table 3.4 Continued.

Area	Characteristics	Sample view and source
Jingyuan, Baiyin, Gaolan, and Yuzhong	The silt content is high. The particles are closely packed, and several trellis pores exist.	(Wang & Deng, 2013)
Jingtai, Xiji, and Haiyuan	The particles are large, distinguishable, and densely packed. The cementation is poor.	(Deng, 2009)
Qingbaishi	The particles are mostly individual particles. The cementation is weak. The particles are loosely packed, and the trellis pores are large.	(Deng, 2009)

(continued)

Table 3.4 Continued.

Area	Characteristics	Sample view and source
Huining	The particles are mostly loosely packed individual particles. The cementation is poor. Intergranular pores abound.	 (Deng, 2009)
Pingliang and Xifeng	The particles are covered with clay. The cementation is moderate. Trellis pores are numerous.	 (Deng et al., 2010)
Luochuan	The particles are mostly loosely packed and moderately cemented into aggregates. Intergranular pores are generally large.	 (Liu, 2008)

(continued)

Table 3.4 Continued.

Area	Characteristics	Sample view and source
Kelan	The particles are loosely to moderately packed individual particles and aggregates. The cementation is moderate. Trellis and mosaic pores exist.	(Wang, 2010)
Taiyuan	The particles are mostly loosely packed individual particles and aggregates. The cementation is poor. Trellis pores are well developed.	(Wang, 2010)

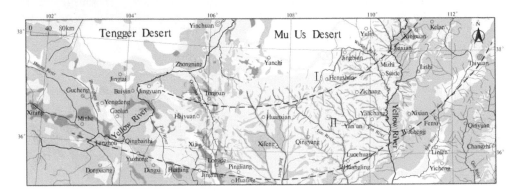

Figure 3.10 Silty loess zone.

Table 3.5 SEM structures of the Malan loess in certain areas of the clayey loess zone.

Area	Characteristics	Sample view and source
Lintao, Tongwei, and Jingning	The particles are coated with clay. Trellis pores are well developed.	(Deng et al., 2010)
Weiyuan	Most of the particles are smaller than 50 μm, moderately packed, and moderately cemented.	(Deng, 2009)
Tianshui	The aggregates form clusters, with poor cementation between aggregate clusters. Trellis pores are well developed.	(Deng, 2009)

(continued)

Table 3.5 Continued.

Area	Characteristics	Sample view and source
Baoji, Weinan, and Pucheng	The particles are mostly fine silt and clay. The particles with average diameters of 18–35 μm fill the voids between coarse particles.	(Deng et al., 2010)
Xi'an	The particles are mainly individual particles and several aggregates composed of small particles. The particles are moderately to well packed. Calcium carbonate is the cementing agent. Trellis and mosaic pores abound.	(Wang, 2010)
Huayin	The particles are mainly moderately packed individual particles, which are poorly cemented with clay and other agents. Trellis pores are frequent.	(Wang, 2008)

(continued)

Table 3.5 Continued.

Area	Characteristics	Sample view and source
Tongguan	Closely packed aggregate clusters are predominant, with occasional coarse particles. The cementation is strong. A few large pores exist.	 ×300 100μm (Wang, 2008)
Yuncheng	The loess contains subangular and subrounded moderately packed individual particles. Abundant large trellis pores and several small mosaic pores exist.	 ×300 200μm (Wang, 2010)
Linfen	Most of the particles form loosely packed and moderately cemented aggregates. The trellis pores are well developed.	 ×500 50μm (Wang & Deng, 2013)

(continued)

Table 3.5 Continued.

Area	Characteristics	Sample view and source
Changzhi	The particles are very closely packed and well cemented with clay and salt crystals. A few large trellis pores and numerous small intergranular pores exist.	×500 50μm (Yuan, 2008)
Yanshi	Nearly all particles form aggregate clusters, which are closely packed and well cemented. Uniformly distributed small trellis pores abound.	×305 100μm (Wang, 2008)

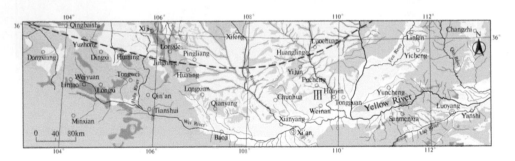

Figure 3.11 Clayey loess zone.

3.5 SUMMARY

The grain size composition of loess geographically varies; consequently, the microstructural characteristics of the Malan loess change roughly consistently from northwest

Table 3.6 Regional variation of the microstructure of the Malan loess.

Zones	Predominant particle types	Roundness	Contact form	Cementation	Intergranular pores (%)
Sandy loess	Granular	Angular	Contact	Thin film	4.95–6.15
Silty loess	Granular, Granular–aggregate	Subangular	Contact–cement	Interstitial	2.74–4.95
Clayey loess	Aggregate	Subrounded, rounded	Cement	Matrix	<2.5

to southeast across the Loess Plateau (Table 3.6) (Liu & Zhang, 1962; Liu, 1965, 1966):

a)   Regional changes in individual particles: Very few particles with diameters larger than 0.1 mm are formed in the Malan loess. The diameters of the particles generally range from 0.02 mm to 0.04 mm. The sand content decreases from northwest to southeast over the loess depositional areas of China. Similarly, the roundness of particles gradually increases. (Zhu, 1963; Wang & Bao, 1964).

b)   Regional changes in cementation: The degree of cementation in the Malan loess increases from northwest to southeast, as the form of cementation gradually changes from coating to matrix type (Zhang, 1964; Li, 1997). For example, in the vicinities of the counties of Hengshan and Jingbian (Fig. 3.7), clay coats the particles. The thickness of the coating film is less than 0.01 mm, and linkages between the coating films are rare. In the vicinities of Yan'an City (Shaanxi) and Lishi District (Shanxi) (Fig. 3.7), the particles are cemented together firmly. In the clayey loess zone where cementation mainly forms aggregates, the cement content is significantly greater than in other zones. The change in the form of cementation is related to the regional variation in clay content.

c)   Regional variation in intergranular pores: From northwest to southeast across the Loess Plateau, the particles of the Malan loess become finer, the cement content increases, and the diameters and abundance of the intergranular pores decrease. In sandy loess, the maximum diameter of the intergranular pores is approximately 0.05 mm. The intergranular pores form a weak network and account for 4.95%–6.15% of the loess volume. In silty loess, the diameters of the intergranular pores usually range from 0.03 mm to 0.05 mm; thus, they are smaller than those in sandy loess. The intergranular pores generally account for approximately 2.74% of the loess volume. In clayey loess, the diameters of the intergranular pores are less than 0.03 mm, and their content is less than 2.5%. Other types of large pores abound, and their walls are covered by carbonate minerals, which form a relatively firm structure.

In the northwestern region of the Loess Plateau, where the climate is dry, calcium carbonate does not leach out and forms effective cementation at contacts, particularly at point contacts. During the deposition process, a porous structure is formed because of the calcium carbonate. In the southeast region, the climate is humid, and calcium carbonate leaches out. The aggregates are combined into clusters because of the high clay content. The particles are tightly packed but mainly linked by surface cementation. In general, the loess microstructure gradually transforms from granular–trellis–contact structure to aggregate structure, toward the southeast; thus, the loess increases in strength and decreases in collapse potential (Gao, 1980b, 1984; Lei, 1989; Wang, 2010).

REFERENCES

Cai, W.L. (2014) *Interaction between the skeleton particles loess* (In Chinese). PhD Thesis. Chang'an University, Xi'an.

Deng, J. (2009) *The regional environment of forming microstructure of loess and mechanism of its seismic subsidence* (In Chinese). PhD Thesis. Lanzhou University, Lanzhou.

Deng, J., Wang, L.M., Zhang, Z.Z. & Bing, H. (2010) Microstructure characteristics and forming environment of late quaternary period loess in the loess plateau of China. *Environment Earth Sciences*, 59(8), 1807–1817.

Feng, L.C. & Zheng, Y.W. (1982) *The Collapsible Loess in China* (In Chinese). Beijing, China Railway Publishing House.

Gao, G.R. (1980a) The classification of loess microstructure and the collapsibility of loess (In Chinese). *Science China Mathematica*, 10(12), 1203–1208.

Gao, G.R. (1980b) The microstructure of Chinese loess (In Chinese). *Chinese Science Bulletin*, 20, 945–948.

Gao, G.R. (1984) Microstructure of loess soil in China relative to geographic and geologic environment (In Chinese). *Acta Geologica Sinica*, 03, 265–272.

Lei, X.Y. (1983) Type of the loess microtextures in Xi'an district (In Chinese). *Journal of Northwest University*, (4), 54–65.

Lei X.Y. (1987) The types of loess pores in China and their relationship with collapsibility. *Scientia Sinica*, 11, 1398–1408.

Lei, X.Y. (1988) Classification and engineering properties of loess pores in China (In Chinese). In: The Geological Society of China geological engineering Specialized Committee (ed.) *The third national conference of engineering geology papers, 2–7, Dec, 1988, Chengdu, China*. Chengdu, Chengdu University of Science and Technology Press. pp. 15–23.

Lei, X.Y. (1989) The relationship between the microstructure types and the indices of physic-mechanical properties in loess of China (In Chinese). *Acta Geologica Sinica*, (2), 182–191.

Li, L., Wang, L.M. & Liu, X. (2005) Research on loess micro-structure in the magistoseismic area (In Chinese). *Earthquake research in China*, 21(3), 369–377.

Li, P. (1997) *Study on Longdong loess strata and collapsibility* (In Chinese). PhD Thesis. Xi'an Polytechnic University, Xi'an.

Liu, H.S. (2008) *Study on the mechanical properties of loess considering stress history and depositional environment* (In Chinese). PhD Thesis. Chang'an University, Xi'an.

Liu, T.S. (1965) *The Deposition of Loess in China* (In Chinese). Beijing, Science Press.

Liu, T.S. (1966) *Composition and Texture of Loess* (In Chinese). Beijing, Science Press.

Liu, T.S. (1985) *Loess and Environment* (In Chinese). Beijing, Science Press.

Liu, T.S. & Zhang, Z.H. (1962) Loess in China (In Chinese). *Acta Geologica Sinica*, 42(1), 1–18.

МУстафасВ, A.A. (1984) *Calculation of Collapsible Loess Foundation*. Beijing, China Water & Power Press.

Wang, L. & Deng, J. (2013) The loess microstructure kinds and its seismic subsidence. *Natural Science*, 05(7), 792–795.

Wang, M. (2010) *Study on structure of collapsible loess in China* (In Chinese). PhD Thesis. Taiyuan University of Technology, Taiyuan.

Wang, Q., Zhang, Q.Y. & Tang, D.X. (1992) Effect of free oxides on the engineering geological properties of soils (In Chinese). In: The Geological Society of China Geological Engineering Specialized Committee (ed.) *The fourth national conference of engineering geology papers (2), 23 Oct 1992, Beijing, China.* Beijing, China Ocean Press. pp. 992–999.

Wang, T.M. & Bao, Y.Y. (1964) *Grain Size Analysis of Loess in the Middle of the Yellow River* (In Chinese). Beijing, Science Press.

Wang, X.J. (2008) *Study on the great and important engineering geological problems for high-speed railway construction in loess area (Illustrated with Zhengzhou-Xi'an passenger dedicated line)* (In Chinese). PhD Thesis. Lanzhou University, Lanzhou.

Wang, Y.Y. (1982) *Loess and Quaternary Geology* (In Chinese). Shanxi, Shanxi people's publishing House.

Wang, Y.Y. & Teng, Z.H. (1982) The Microstructure of Chinese loess and its changes in time and area (In Chinese). *Chinese Science Bulletin*, (2), 102–105.

Xing, J.X. (2004) *The analysis and research on the influence factors of loess collapsibility* (In Chinese). PhD Thesis. Chang'an University, Xi'an.

Xing, Y.D., Zhu, F.S. & Wang, C.M. (2008) Physical components and microstructure characteristics of loess in west of Liaoning (In Chinese). *Geotechnical Engineering Technique*, 03, 155–159.

Yuan, H. (2008) *The experimental and microstructural study on collapsibility of loess* (In Chinese). PhD Thesis. Taiyuan University of Technology, Taiyuan.

Zhang, Y.S. & Qu, Y.X. (2005) Cements of sand loess and their cementation in north Shaanxi and west Shanxi (In Chinese). *Journal of Engineering Geology*, 01, 18–28.

Zhang, Z.H. (1964) Study on the microstructure of loessial soil in China (In Chinese). *Acta Geologica Sinica*, 44(3), 357–369.

Zhu, H.Z. (1963) Some characteristics of the particles and structures of Malan loess in the middle of the Yellow River (In Chinese). *Scientia Geologica Sinica*, 9(2), 88–100.

Zhu, H.Z. (1964) *The Phenomenon of Grain Size Variation of Loess in the Middle of the Yellow River and Its Interpretation* (In Chinese). Beijing, Science Press.

Zhu, H.Z. (1965) Microstructure of loess and the light clay in buried soil (In Chinese). *Quaternary Sciences*, 4(1), 62–76.

Chapter 4

Physical and mechanical properties

4.1 CLASSIFICATION OF LOESS

Although different classification schemes of loess strata were proposed according to petrological characteristics, paleontological fossils, erosion surface, and other characteristics (Table 4.1), the one by Liu (1985) is the most often used in China. In the classification scheme, the Quaternary loess is divided into four main parts from top to bottom of loess strata, namely, the Holocene loess, the Late Pleistocene Malan loess, the Middle Pleistocene Lishi loess, and the Early Pleistocene Wucheng loess.

Holocene loess (Q_4), also known as the modern secondary loess, began to form about 7500 years ago. Its thickness is generally less than 2.00 m; in some areas, the thickness can exceed 10.00 m. Holocene loess is yellow or brown yellow. In some parts of Northwest China, a layer of Heilu soil exists at the bottom of Holocene loess. The thickness of Heilu soil is generally 0.10–0.50 m, and its color is dark gray or brownish gray. Generally, Holocene loess has low dry density, high void ratio, large compressibility, and high permeability. Under dry conditions, the structure of Holocene loess is strong but it can be damaged once it is saturated with water. Furthermore, the cohesion decreases rapidly and demonstrates a wide variation range. Therefore, Holocene loess has strong collapsibility.

The Late Pleistocene Malan loess (Q_3) was formed in the period between 100 thousand and 7500 years ago. Its thickness is generally 5.00–30.00 m, with the thickest portion located in the middle of Gansu Province, China. The Malan loess located between the Heilu soil and the first paleosol layer is grayish yellow or orange. It is loose and porous, with well-developed vertical joints and uniform texture. This layer contains high carbonate content and calcareous concretions. The physical and mechanical properties of the Late Pleistocene Malan loess are similar to those of Holocene Loess.

Lishi loess (Q_2) in the middle Pleistocene, was formed in the period between 1.15 million and 0.1 million years ago. The Q_2 loess is yellow or light brown yellow. The thickness of Q_2 loess ranges from 50 to 200 m with totally 12 layers of paleosol. The loess can be divided into two sections that are bounded by the top of a reddish brown paleosol layer about 5 meters thick. Lishi loess has a dense texture, a high density, a small compressibility and a small permeability. It has no or slight collapsibility under high stress and therefore it is normally used as bearing stratum for foundations.

The Early Pleistocene Wucheng loess (Q_1), called the stone loess (lithified or indulated loess), was formed in the period between 2.4 million and 1.15 million years ago.

Table 4.1 Classification scheme of loess strata (Liu, 1985).

Scheme Age	Richthofen (1877)	Andersson (1923)	de Chardin & Yang (1930)	Liu (1959)	Liu & Zhang (1962)	Liu & Wang (1964)	Liu (1985)
Holocene	Q_4 Loess	Secondary loess				Loess-like rock	Holocene loess
Late Pleistocene	Q_3	Malan loess (primary loess)	Malan loess	New loess	Malan loess	Malan loess Qianxian County group	Malan loess
Late Middle Pleistocene	Q_2^2		Red soil	Old loess	Upper Lishi loess Dingcun group	Upper Lishi loess	Upper Lishi loess
Early Middle Pleistocene	Q_2^1				Lower Lishi loess	Lower Lishi loess Shanxian group	Lower Lishi loess
Early Pleistocene	Q_1				Wucheng loess	Wucheng loess	Wucheng loess

The loess is reddish yellow, and its thickness ranges from 30 m to 80 m. The Wucheng loess is denser, stronger, less compressible, and less permeable than the Lishi loess; furthermore, the Wucheng loess is non-collapsible (Liu & Zhang, 1962; Qian & Wang, 1985; Wang et al., 1994; Cao, 2005).

4.2 PHYSICAL AND MECHANICAL PROPERTIES

The extensively studied physical properties of loess include particle composition, dry density, porosity, specific gravity, moisture content, plastic limit, liquid limit, plasticity index, and liquidity index. The main mechanical properties, which were widely investigated in the literature, include shear strength, compressive modulus, compressibility coefficient, and collapsibility coefficient. The physical properties, mineral composition, and microstructure of loess affect its mechanical properties. In the following sections, the dry density, natural moisture content, plasticity index, liquidity index, compressibility coefficient, compressive modulus, and collapsibility coefficient of loess are discussed.

4.2.1 Dry density

The dry density of loess is the ratio of the mass of solid particles to the total volume of loess. This parameter serves as a physical index for expressing the bearing capacity

Table 4.2 Dry densities of loess deposits in different sedimentary ages (g/cm³) (Statistic analysis on data from [1–8], [12–19], [21–23], [25], [27–29], [31–39], [41–46]).

Stratum	Maximum value	Minimum value	Average value	Standard deviation
Q_4	1.85	1.08	1.296	0.138
Q_3	1.9	1.19	1.40	0.17
Q_2	1.73	1.47	1.561	0.08
Q_1	1.886	1.55	1.697	0.0997

and the degree of soil compactness of cohesionless soil. Dry density is affected by numerous factors, such as the buried depth, the structure of loess grain, the degree of cementation, the moisture content, and the particle composition. Generally, the higher the soil dry density is, the more compact the soil particles are, the lower the compressibility is, the higher the bearing capacity of the foundation is, and the better the engineering properties of the soil are.

As shown in Table 4.2, the greater the sedimentary age is, the higher the dry density of the loess is. Compared with the Holocene loess and Malan loess with small buried depths, the Wucheng loess and Lishi loess present smaller standard deviations of dry densities. These statistics indicate that the dry densities of old loess are more concentrated. The large buried depths of the Wucheng loess and Lishi loess led to their relatively high porosity and dry density as well as low dispersion. By contrast, the Malan loess and Holocene loess were formed relatively late, thus, they can be easily disturbed by external factors, such as human activities. The soil properties are unstable, and the density is low.

4.2.2 Natural moisture content

The natural moisture content of loess is the ratio between the mass of soil water and that of soil particles under natural conditions. The change in the moisture content of loess is not only related to the external humidity but also to the content of clay and organic matter in loess. When the external humidity conditions are the same, higher contents of clay and organic matter in loess indicate greater amount of absorbed water, and, consequently, higher moisture content of the loess.

Table 4.3 shows that the moisture contents of the Lishi loess and Wucheng loess are greater than those of the Holocene loess and Malan loess. These statistics are attributed to the large buried depths of the Wucheng loess and Lishi loess. Their large buried depths can accommodate certain amounts of rainwater and groundwater, which do not easily evaporate compared with those in shallow loess.

In the same profile, the moisture content of loess fluctuates from top to bottom. The peak is located in the paleosol layers, and the trough is recorded in the loess layers. In the loess layers with calcareous concretions, the moisture content initially decreases and subsequently increases with the depth. The peak is at the top of the calcareous concretion layer because of the – water-resistance effect of the calcareous concretion layers. In the paleosol layers, the moisture content initially increases and subsequently

Table 4.3 Natural moisture contents of loess deposits of different sedimentary ages (%) (Statistic analysis on data from [1–46]).

Stratum	Maximum value	Minimum value	Average value	Standard deviation
Q_4	32.4	4.7	12.02	5.83
Q_3	33.6	3	12.50	6.69
Q_2	23.6	2.8	15.91	5.09
Q_1	28.1	6.3	15.35	7.91

Table 4.4 Plasticity index values of loess deposits in different sedimentary ages (Statistic analysis on data from [1–7], [10–12], [15], [20], [22–26], [28–29], [31–32], [36–40], [42], [44], [46]).

Stratum	Maximum value	Minimum value	Average value	Standard deviation
Q_4	18.0	4.0	9.65	2.53
Q_3	14.1	2.3	9.45	3.14
Q_2	16	7.1	10.07	3.31
Q_1	13.95	6.45	10.18	2.34

decreases with the depth. The peak value appears in the middle and lower parts of the paleosol layers, as the clayization is the strongest in these parts (Xu & Zhao, 2002).

4.2.3 Plasticity index

Plasticity index (I_P) is the range of in the soil moisture content in the plastic state. This index is affected by the particle sizes and the mineral composition of the loess. The finer the particle sizes are, the larger the specific surface area is, the higher the clay content is, the lower the permeability is, and ultimately the greater the moisture content. The plasticity index of loess is one of the measures of the water-physical property. This index can reflect not only the soil structure (e.g., particle composition and mineral composition) but also the collapsibility of the loess and the bearing capacity of the foundation (Zhu, 1995).

Table 4.4 shows that the plasticity index of loess increases with the sedimentary age. The plasticity index is directly proportional to the clay content (Fig. 4.1), and the clay content increases relatively with the sedimentary age (Table 4.5); consequently, the plasticity index of old loess is greater than that of new loess.

4.2.4 Liquidity index

Liquidity index (I_L) is a parameter for measuring the soft and hard states of clayey soil and silty soil. It is the ratio of the difference between natural moisture content and plastic limit to the plasticity index. The state of loess can be classified as solid, semi-solid, plastic, or flow according to the liquidity index. A liquidity index lower than 0 indicates high soil hardness, that is, the loess is in a solid state. A liquidity index

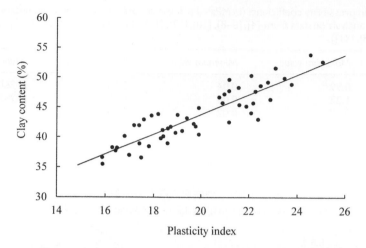

Figure 4.1 Relation fitting curve of plasticity index and clay content (Zhang & Chen, 2013).

Table 4.5 Particle compositions of the loess deposits in Northwest China (Cao, 2005).

	Particle size (mm)			
	>0.05	0.05–0.01	0.01–0.005	<0.005
Loess type	Content (%)			
Malan loess	13.3	61.5	7.6	17.6
Upper Lishi loess	11.9	59.3	8.6	20.2
Lower Lishi loess	10.5	56.8	6.9	25.8
Wucheng loess	7.9	56.1	9.8	26.2

greater than 1 shows that the soil moisture content is high, and the loess is in a flow state. A liquidity index between 0 and 1 indicates that the soil is in a plastic state. The higher the I_L value is, the softer the loess is, and vice versa.

4.2.5 Compressibility coefficient

The compressibility of loess is the property of loess volume reduction under pressure. It is also one of the basic concepts of the law of porosity. Compressibility coefficient, which is a physical quantity that describes the compressibility of loess, is equal to the slope of the secant at a certain pressure range of e–p curves obtained from a compression test. A greater compressibility coefficient indicates higher compressibility of the loess.

Table 4.6 shows that greater sedimentary age indicates lower compressibility coefficient, and the less dispersion. This result is because of the fact loess with greater

Table 4.6 Compressibility coefficients (in MPa^{-1}) of loess deposits in different sedimentary ages (Statistic analysis on data from [4], [6–8], [10], [12], [14], [17–19], [22], [25], [28–29], [31], [33–34], [38], [41]).

Stratum	Maximum value	Minimum value	Average value	Standard deviation
Q_4	0.99	0.04	0.399	0.268
Q_3	1.22	0.04	0.329	0.297
Q_2	0.27	0.02	0.149	0.071
Q_1	0.44	0.0022	0.110	0.144

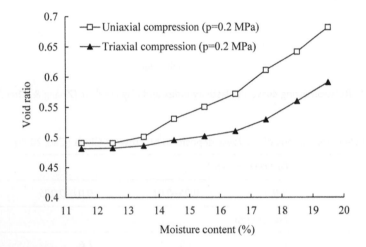

Figure 4.2 Relationship between compressibility and moisture content (Jing & Zhang, 2004).

depth presents stronger the compaction and the lower porosity. The shallow depth of new loess renders the loess vulnerable to external disturbance and subject to severe weathering and denudation. As a result, new loess presents a loose structure, high compressibility, strong regional feature, and large data discreteness.

Jing and Zhang (2004) found that when the moisture content of loess is less than the optimum moisture content (13.5%), its compressibility does not change remarkably with the moisture content. However, when the moisture content is more than the optimum moisture content, the loess porosity ratio increases significantly with an increase in the moisture content; thus, the compressibility decreases with an increase in the moisture content (Fig. 4.2).

4.2.6 Compressive modulus

The compressive modulus of loess is another parameter indicating the compressibility of loess. It is the ratio of the vertical stress to the strain of the loess under the condition of lateral confinement. In terms of the compressibility of the loess, smaller compressive modulus indicates greater compressibility coefficient, higher compressibility, and the

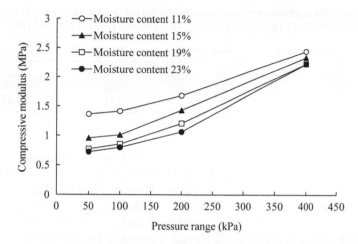

Figure 4.3 Variation in compressive modulus with water content (Yuan, 2015).

more loosely packed loess particles. In engineering, compressive modulus is usually used to determine the degree of compression deformation.

The softening effect of water is an important factor affecting the compressibility of loess. By conducting experiments, Yuan (2015) concluded that with the same moisture content, the compressive modulus of undisturbed loess increases significantly with an increase in the pressure range. In the same pressure range, the compressive modulus of undisturbed loess decreases with an increase in moisture content (Fig. 4.3).

4.2.7 Collapsibility coefficient

The collapsibility of loess refers to the property of the soil structure being rapidly destroyed and the loess suddenly sinking when soaked in water under its weight or an external load. The collapsibility coefficient of loess is a mechanical parameter for evaluating the collapsibility of loess, and it refers to the ratio of the height difference of a loess sample to its original height before and after soaking at a certain pressure. The collapsibility coefficient has two types, namely, self-weight collapsibility coefficient and nonself-weight collapsibility coefficient. This coefficient can be determined by conducting an indoor test. Generally speaking, if the collapsibility coefficient is lower than 0.015, then the loess exhibits no collapsibility. If the collapsibility coefficient is 0.015–0.03, then the loess has slight collapsibility. If the collapsibility coefficient is 0.03–0.07, then the collapsibility is moderate. If the collapsibility coefficient is greater than 0.07, then the loess has strong collapsibility. The lower the collapsibility coefficient, the worse the collapsibility is, and the better the physical and mechanical properties are.

Collapsible loess must meet two conditions, namely, collapsible space and a certain amount of connection with which loess strength will decrease with water (Wang, 2007). Table 4.7 shows that as the sedimentary age increases, the collapsibility coefficient of loess decreases. The changing pattern can be roughly understood from the law of stratigraphic sequence. The porosity of the old loess decreases after millions of

Table 4.7 Collapsibility coefficients of loess deposits in different sedimentary ages (%) (Statistic analysis on data from [6–7], [9–11], [18], [20–21], [23], [27–28], [30–32], [34–36], [38], [41–42], [44]).

Stratum	Maximum value	Minimum value	Average value	Standard deviation
Q_4	12.99	0.61	4.70	3.29
Q_3	15.1	0.1	4.52	3.70
Q_2	4.6	0.2	1.41	1.53

years of compaction because the deposition of old loess is earlier than that of new loess; consequently, the old loess is dense. Therefore, porosity of the old loess is not a prerequisite for the occurrence of a collapse. The new loess has high porosity, loose structure, uneven texture, and high compressibility because it has short formation time and has been subjected to severe weathering and erosion. The new loess contains numerous hydrophilic minerals, and the salt content in the local area is high. These conditions are favorable to the infiltration of water, and they may lead to serious collapsibility of new loess.

REFERENCES

[1] An, Y.F., Liang, Q.G., Zhao, L. & Zhang, Y.J. (2011) Anisotropy of mechanical properties of Q_4 loess in Lanzhou (In Chinese). *Journal of Lanzhou Jiaotong University*, 30(1), 90–96.

[2] Chen, C.L., He, J.F. & Yang, P. (2007) Constitutive relationship of intact loess considering structural effect (In Chinese). *Rock and Soil Mechanics*, 28(11), 2284–2290.

[3] Chen, H. (2011) Multi factor experimental study on static characteristics of saturated loess (In Chinese). *Railway Survey and Design*, (5), 45–49.

[4] Duan, Z.F. & Xu, H.A. (1992) Engineering geological characteristics and distribution law of soft loess in Xi'an area (In Chinese). *Journal of Earth Sciences and Environment*, (1), 64–70.

[5] Guo, J., Luo, Y.S., Guo, H. & Fu, Z.Y. (2010) Experimental study on structural characteristics of loess in different regions (In Chinese). *Bulletin of Soil and Water Conservation*, 30(1), 89–92.

[6] Hu, Z.Q. & Shen, Z.J. (2000) Study on the structure of unsaturated loess (In Chinese). *Chinese Journal of Rock Mechanics and Engineering*, 19(6), 775–779.

[7] Hu, Z.Q., Shen Z.J. & Xie, D.Y. (2004) Deformation properties of structural loess (In Chinese). *Chinese Journal of Rock Mechanics and Engineering*, 23(24), 4142–4146.

[8] Hua, Z.M., Zhang S.A., Zhang, E.X. & Shen, Q.W. (2010) Analysis of the foundation design of high rise buildings in Longdong Loess Plateau (In Chinese). *Geotechnical Investigation & Surveying*, S1, 267–277.

[9] Jia, H., He, Y.L. & Yang, S.X. (2011) Study on the influence of joints on the failure modes and mechanical properties of Loess (In Chinese). *China Rural Water and Hydropower*, (11), 77–81.

[10] Jia, L. (2015) Study on variability of physical and mechanical properties of Loess (In Chinese). *Northern Communications*, (9), 44–46.

[11] Jiang, M.J., Hu, H.J., Peng, J.B. & Yang, Q.J. (2012) Pore changes of loess before and after stress path tests and their links with mechanical behaviors (In Chinese). *Chinese Journal of Geotechnical Engineering*, 34(8), 1369–1378.

[12] Lai, T.W. & He, B. (2003) Variability and correlation analysis of physical and mechanical properties of Loess (In Chinese). *Journal of Lanzhou Jiaotong University*, 22 (6), 140–140.

[13] Lei, S.Y. & Tang, W.D. (2004) Analysis of microstructure change for loess in the process of loading and collapse with CT scanning (In Chinese). *Chinese Journal of Rock Mechanics and Engineering*, 23(24), 4166–4169.

[14] Li, B.X. & Li, Y.J. (2003) Engineering geological characteristics of Malan loess in Lanzhou (In Chinese). *Journal of Gansu Sciences*, 15(3), 31–34.

[15] Li, T.L., Zheng, S.Y., Deng, H.K. & Zhao, J.L. (2004) Collection and Disposal of Earlier Pavement Runoff (In Chinese). *Highway*, (10), 37–41.

[16] Li, W.T. (2015) Experimental study on Collapsibility of loess in a Hydropower Station (In Chinese). *Shaanxi Water Resources*, (1), 111–113.

[17] Li, Y.L. (1995) Study on the basic characteristics and engineering geological properties of loess in West Henan (In Chinese). *Journal of North China University of Water Resources and Electric Power (Natural Science Edition)*, (2), 31–37.

[18] Li, Z.Q., Yu, W.L., Fan, L.F., Fu, L., Hu, R.L., Lin, D.J. & Wang, Y.P. (2011) Experimental research on strength characteristics and engineering treatment of improved loess soil (In Chinese). *Journal of Engineering Geology*, 19(1), 116–121.

[19] Liu, W.F. & Hu, A.P. (2008) The influence of rainwater permeation in loess plateau on its engineering geology (In Chinese). *China Building Materials Science & Technology*, 17(1), 69–72.

[20] Ou, Z.Y., Deng, J.C., Cai, J.H., Duan, Z.Q. & Xu, G.T. (2008) Engineering geology of loess channel of Shandong section of eastern route of the south-to-North water diversion (In Chinese). *South-to-North Transfers and Water Science & Technology*, 6(1), 92–94.

[21] Tian, K.L., Wang, P. & Zhang, H.L. (2013) Discussion on stress-strain relation of intact loess considering soil structure (In Chinese). *Rock and Soil Mechanics*, 34(7), 1893–1898.

[22] Tong, X.J. & Liu, G.M. (2013) Study on physical and mechanical properties and treatment methods of collapsible loess in Changqing District of Ji'nan (In Chinese). *Site Investigation Science and Technology*, (2), 39–42.

[23] Wang, J. (2008) Discussion on loess vulnerability and seismic loess landslides (In Chinese). *Journal of Gansu Sciences*, 20(2), 36–40.

[24] Wang, T. & Jiang, X.F. (2015) Origin Analysis and Engineering and Geological Characteristics of Loess in Ji'nan Area (In Chinese). *Shandong Land and Resources*, 31 (12), 46–49.

[25] Wang, X.J., Mu, N.S. & Wu, Q. (2014) Statistical characteristics of physico-mechanical indices of the new loess (In Chinese). *Highway Engineering*, (1), 88–93.

[26] Wang, X.M., Chen, S.X. & Cheng, C.B. (2013) Experimental study on physico-mechanical characteristics of undisturbed loess soaked in acid solution (In Chinese). *Chinese Journal of Geotechnical Engineering*, 35(9), 1619–1626.

[27] Wang, Z. (2011) Experimental study on consolidation of collapsible loess subgrades by impact rolling compaction on Shijiazhuang-Taiyuan passenger-dedicated line (In Chinese). *Railway Standard Design*, (9), 30–32.

[28] Wang, Z.Q. (2005) Engineering geology of saturated loess in Yintao water supply project, Gansu (In Chinese). *Journal of Engineering Geology*, 13(4), 471–476.

[29] Wei, F., Cao, Z.L. & Ning, W.L. (1996) Discussion on collapsible loess in Taiyuan area (In Chinese). *West-china Exploration Engineering*, (S1), 15–17.

[30] Wei, J.W. (2009) Discussion on the change of mechanical parameters of collapsible loess under humidification condition (In Chinese). *Site Investigation Science and Technology*, (5), 41–44.

[31] Wu, Y.P. & Zhao, C.X. (2001) The engineering geological properties of Fulongping District Loess in Lanzhou City (In Chinese). *Journal of Gansu Sciences*, 13(4), 41–43.

[32] Xia, W.M. & Guo, Z.Y. (2004) Elasto plastic softening constitutive model of Q_1 loess (In Chinese). *Journal of Xi'an University of Technology*, 20(3), 241–244.

[33] Xie, F.C. & Jiang, Z.F. (1987) Basic characteristics of loess in West Henan (In Chinese). *Henan Geology*, (3), 44–50.

[34] Xing, Y.D., Wang C.M., Zhang, L.X. & Kuang, S.H. (2008) Subgrade treatment effects for collapsible loess subgrade of Fuxin-Chaoyang free way in west of Liaoning province (In Chinese). *Journal of Jilin University (Earth Science Edition)*, 38(1), 98–104.

[35] Yan, B. & Xu, Z.J. (2009) Subgrade loess collapsibility before and after low level tamping (In Chinese). *Journal of Chang'an University (Natural Science Edition)*, (4), 25–29.

[36] Yang, F., Chang, W., Wang, F.W. & Li, T.L. (2014) Motion simulation of rapid long run-out loess landslide at Dongfeng in Jingyang, Shaanxi (In Chinese). *Journal of Engineering Geology*, 22(5), 890–895.

[37] Yin, G.H., Wang, L.M., Yuan, Z.X., Liu, H.M., Wang, P. & Wu, G.D. (2009) Physical properties, dynamic characteristics and landslide of loess in Xinjiang, Yili (In Chinese). *Arid Land Geography*, 32(6), 899–905.

[38] Yu, Y.H. & Yang, X.H. (2003) Experimental study on mechanical properties of cement loess (In Chinese). *Journal of Chang'an University (Natural Science Edition)*, 23(6), 29–32.

[39] Yuan, K.F., Li, Z.J. & J, L. (2013) Correlation between compression modulus and collapsibility of Loess (In Chinese). *Journal of Liaoning Technical University* (Natural Science), 32(11), 1480–1483.

[40] Zhang, B.L., Gao, L.C., Liu, H.W. & Zhu, J.B. (2009) Study on Influencing Factors of collapsible loess in Handan area (In Chinese). *Geotechnical Investigation & Surveying*, (S2), 17–19.

[41] Zhang, S.A. (1994) Engineering characteristics and nonlinear parameters determination of the old loess in Jiuzhoutai section (In Chinese). *Hydrogeology and Engineering Geology*, (5), 22–25.

[42] Zhang, Z.L. (2003) Engineering characteristics of collapsible loess in Tianshui (In Chinese). *Mineral Exploration*, 6(9), 71–73.

[43] Zhou, Q., Zhao, F.Z. & Zhang, H.L. (2006) Effect of compaction degree and water content on mechanical properties of compacted loess (In Chinese). *Highway*, (1), 67–70.

[44] Zhou, X.Y. & Luo, Y.S. (2007) Influence of stress path on mechanical properties of saturated loess drainage test (In Chinese). *Chinese Journal of Underground Space and Engineering*, 06, 1064–1068.

[45] Zhou, Z.J., Yang, H.F., Geng, N. & Ye, W.J. (2013) Influence of freezing speed on physical and mechanical properties of freezing-thawing loess (In Chinese). *Journal of Traffic and Transportation Engineering*, 13(4), 20–25.

[46] Zhu, L.F., Hu, W., Zhang, M.S., Tang, Y.M., Bi, J.B. & Ma, J.Q. (2013) An analysis of the soil mechanical properties involved in loess landslides in Heifangtai, Gansu Province (In Chinese). *Geological Bulletin of China*, 32(6), 881–886.

[47] Andersson, J.G. (1923) Essays on the genozoic of northern China (In Chinese). *Mem. Geol. Surv. China*, Ser. A. 3.

[48] Cao, X.P. (2005) *Study on mechanical properties of loess in Lanzhou area and its application* (In Chinese). Msc Thesis. Lanzhou University, Lanzhou.

[49] de Chardin, P.T. & Yang, Z. (1930). *Preliminary observations on the pre-Loessic and post-Pontian formations in western Shansi and northern Shensi.* Geological survey of China (under the Ministry of agriculture and mines and affiliated with the Academia sinica) and the Section of geology of the National academy of Peiping.

[50] Jing, H.J. & Zhang, B. (2004) Strength law of Loess Subgrade (In Chinese). *Journal of Traffic and Transportation Engineering*, 4 (2), 14–18.

[51] Liu, T.S. (1959) New loess and old loess. *Geology in China*, (5), 24–27.

[52] Liu, T.S. (1985) *Loess and Environment* (In Chinese). Beijing, Science Press.

[53] Liu, T.S. & Wang, K.L. (1964) *Quaternary Geological Problems* (In Chinese). Beijing, Science Press.

[54] Liu, T.S. & Zhang, Z.H. (1962) Loess in China (In Chinese). *Acta Geologica Sinica*, 42 (1), 1–14+106–109.

[55] Qian, H.J. & Wang, J.T. (1985) *Collapsible Loess Foundation* (In Chinese). Beijing, China Architecture & Building Press.

[56] Richthofen, F.V. (1877) China (Vol. 1). *Berlin: Verlag von Dietrich Reimer*, 758.

[57] Wang, J.T., Li, Y.J. & Li, B.X. (1994) Physical characteristics of loess in Lanzhou (In Chinese). *Hydrogeology and Engineering Geology*, (4), 12–17.

[58] Wang, Y.T. (2007) Mechanisms of collapse shown on physical mechanics index of the routine soil test (In Chinese). *Journal of railway engineering society*, 24(3), 1–5.

[59] Xu, Q.X. & Zhao, J.B. (2002) Characteristics of moisture content variation of loess in Xi'an area (In Chinese). *Geoscience*, 16(4), 435–438.

[60] Yuan, K.F. (2015) Experimental study on shear strength and compression modulus of undisturbed loess (In Chinese). *Journal of Yangtze University (Natural Science Edition)*, 12(4), 67–69.

[61] Zhang, M.J. & Chen, L. (2013) Linear regression analysis on clay plasticity index and clay content (In Chinese). *Journal of Zhejiang University of Water Resources and Electric Power*, 25(1), 15–17.

[62] Zhu, M.R. (1995) Classification of plasticity index of Loess Foundation (In Chinese). *Journal of Northwestern Institute of Architectural Engineering*, 6–8.

Chapter 5

Loess permeability

Permeability, which is one of the most important physical indexes of loess, refers to the capability of water to permeate and flow through the pores of loess. Permeability is usually characterized in terms of permeability coefficient. In the loess area, the infiltration of rainfall and irrigation water will lead to weakening of the connection between loess grains, decrease of tensile shear strength, and destruction of soil structure. This action can induce landslides, collapses, and other geological disasters.

The permeability coefficient of loess in the Loess Plateau of China generally ranges from 10^{-8} cm/s to 10^{-4} cm/s. Such large permeability is attributed to the large porosity, small dry density, and dense vertical joints. Individual small pores (<0.004 mm in diameter), which can be easily occupied by bound water in the loess body, and large pores (>0.016 mm in diameter) due to particle overhead are found. These pore characteristics make the permeability of loess stronger than that of fine-grain soils with similar grain size composition (Li, 1991). In addition, loess possesses anisotropic permeability. Generally, the permeability in the vertical direction is much stronger than that in the horizontal direction. This difference can range from several times to dozens of times (Li, 2007; Liu, 2015).

The permeability of loess formed in different geological times also show considerable differences. Compared with Lishi loess (Q_2) and Malan loess (Q_3), Wucheng loess (Q_1) has less distribution and exposure, and is seldom involved in engineering constructions. Studies related to Q_1 loess are also fewer because of the hard texture and the difficulty in making samples for permeability test. The permeability of Lishi loess (Q_2) and Malan loess (Q_3) has received more attention because they are widely distributed on the earth's surface and are often the direct target layer of engineering constructions. The structure of Q_3 loess is loose, and large pores and columnar joints are developed. In the same area, the permeability coefficient of Q_3 loess is more than one order higher than that of Q_2 loess (Lin et al., 2014; Yang et al., 2015).

The factors that affect the permeability of loess include the pore structure of soil, the connection mode between particles, the geographical location and buried depth, as well as the disturbance degree during sampling, and the stress state in the test. Generally, differences in loess permeability are a result of many factors. Changes in any one of the factors will lead to the difference. These factors can be classified into two aspects: internal and external. The internal factors (dry density, water content, etc.) result in the different structural characteristics of the loess, whereas the external factors (such as the confining pressure, permeability, freezing, and thawing cycles)

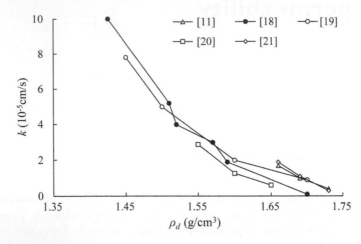

Figure 5.1 Relationship between dry density and permeability coefficient (Data from: [11], [18–21]).

cause the destruction and reorganization of the soil structure. In essence, these factors affect the permeability by changing the pore characteristics of the loess (such as the number, size, and connectivity). The influences of the above factors on the permeability of loess are discussed below.

5.1 DRY DENSITY

The large pores in loess are the main component of the seepage channel, and the small pores and micropores have a lesser effect on the loess's permeability. Loess with small dry density has large permeability coefficient due to the presence of many large pores inside, while loess with large dry density is relatively dense and correspondingly has lower permeability coefficient due to fewer large pores. When the dry density of loess ranges from $1.35 \, \text{g/cm}^3$ to $1.75 \, \text{g/cm}^3$, the permeability coefficient decreases obviously with the increase of dry density. When dry density ranges from $1.35 \, \text{g/cm}^3$ to $1.55 \, \text{g/cm}^3$, the slope of the $k–\rho_d$ curve is very large, indicating that the dry density has a significant impact on the permeability coefficient in this range. When the dry density is greater than $1.60 \, \text{g/cm}^3$, the rate of decrease in permeability coefficient with the decrease of the dry density decreases and tends to be a straight line. Several differences in the variation of permeability coefficient with the dry density are also found for loess samples taken from different regions, but the overall trend is generally consistent (Fig. 5.1).

For example, the shape of particles in Lanzhou loess under SEM at $300 \times$ magnification is irregularly massive and granular (Fig. 5.2). The particles are scattered and arranged in disorder, forming pores with different sizes. In the loess samples with a density of $1.70 \, \text{g/cm}^3$, the difference in particle diameter is large, the particle profile is obvious, the particle grinding roundness is poor, and many large pores are

Figure 5.2   SEM images of loess samples with different density: (a) 1.70 g/cm³; and (b) 1.76 g/cm³ (Zhao & Wang, 2012).

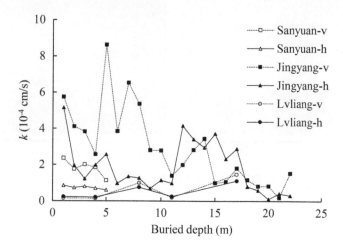

Figure 5.3   Relationship between permeability coefficient and buried depth (Data from: Li, 2007; Liu, 2015).

formed by particle stacking. However, for the samples with a density of 1.76 g/cm³, the difference in particle diameter is small and large pores are significantly fewer.

5.2   BURIED DEPTH

Generally, the lower loess in early geological age has relatively small void ratio and permeability coefficient. This phenomenon is caused by the large compacted pores, which destroyed connecting structure.

Figure 5.3 shows the relationship between permeability coefficient and buried depth of six groups of loess samples taken from Jingyang and Lvliang, Sanyuan. The

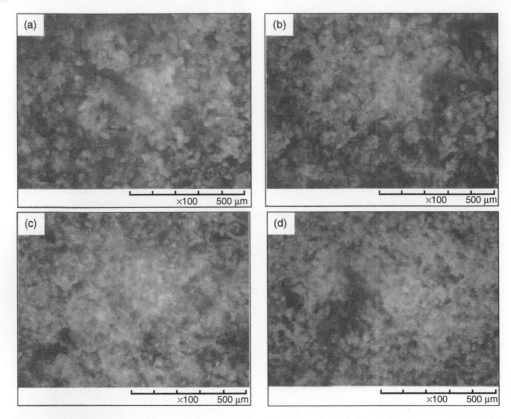

Figure 5.4 SEM images of loess under different depths: (a) 10 m; (b) 20 m; (c) 26 m; and (d) 30 m (Hou et al., 2016).

horizontal (k_h) and vertical (k_v) permeability coefficients of the samples from Sanyuan and Jingyang decrease significantly with the increase in burial depth. The curve with depths of 12 m to 15 m shows a slight fluctuation, because this layer possesses many calcareous nodules in most Jingyang area. Thus, samples are difficult to prepare and side leakage easily occurs during permeability tests. Lvliang area samples were taken from a reservoir loess dam, which is a non-primary loess and therefore unrepresentative. Thus, for this segment, the relationship between the permeability coefficient and the depth shows no obvious rule.

Loess compactness changes significantly with buried depth. As shown in Figure 5.4, the SEM images of loess show the changes in microstructure under depths ranging from 10 m to 30 m. The loess structure changes from loose to dense, and the shape of the particles is gradually condensed from granular to massive. The loess flow channels become relatively blocked because of the increase of loess density and the coagulation of particles. These reasons worsen the loess permeability with the increase in buried depth.

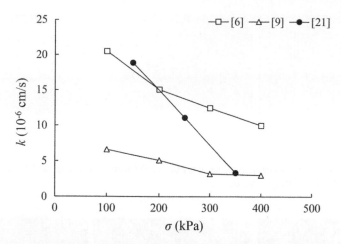

Figure 5.5 Relationship between confining pressure and permeability coefficient (Data from [6], [9], [21]).

5.3 CONFINING PRESSURE

The confining pressure applied in experiments can affect the obtained results of loess permeability. The experimental results show that the permeability coefficient decreases with the increase of confining pressure in ranges of 100 kPa to 400 kPa (Fig. 5.5). The primary loess was used in Wang et al. (2007); the slope of the $k-\sigma$ curve obtained is significantly larger than the other two curves. This indicates that the permeability of primary loess is significantly affected by the confining pressure.

This regularity is also related to the change of loess porosity and structure. In the early stage of the increase of confining pressure, the primary structure of soil is not destroyed and the void ratio decreases continuously. When the confining pressure is increased to a certain extent, not only is the void ratio reduced, but the primary structure of loess is also destroyed. With the rearrangement of loess particles, the fine particles clog the pores and channels, leading to a decrease of permeability coefficient.

Figure 5.6 show the SEM images of the primary loess samples under 0 kPa to 400 kPa consolidation pressure. The loess structure changes obviously under different pressures. Under 0 kPa pressure, the structure is loose, with the presence of many large pores between particles. When the pressure is increased to 200 kPa to 400 kPa, the primary structure of loess samples is destroyed. The arrangement of particles becomes compact, the number of pores decreases, and the pore size also decreases.

Figure 5.7 show the SEM images of the remolded loess samples under different pressures. The black parts with gray values smaller than the set threshold are recognized as pores. In contrast, the white parts with gray values larger than the set threshold are identified as particles. When the pressure is increased from 50 kPa to 400 kPa, the number of particles on the section increases obviously; at the same time, the number of pores decreases.

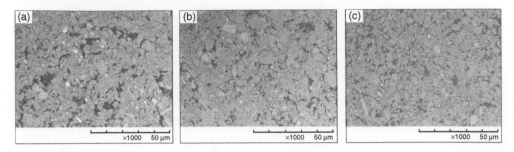

Figure 5.6 SEM images of primary loess samples under different pressures: (a) 0 kPa; (b) 200 kPa; and (c) 400 kPa (Zhang et al., 2015).

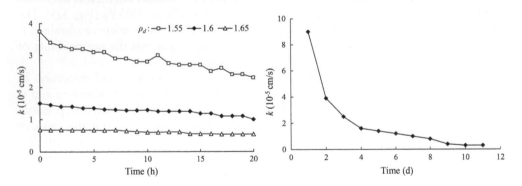

Figure 5.7 SEM images of remolded loess samples under different pressures: (a) 50 kPa; (b) 100 kPa; and (c) 400 kPa (Feng et al., 2013).

Figure 5.8 Relationship between permeation time and permeability coefficient (An, 2011; An et al., 2013).

5.4   PERMEATION TIME

The results of permeability tests at different time scales (hours or days) show that the permeability coefficient decreases with time and tends to stabilize gradually. The initial permeability coefficient values range from several to dozens of times as large as that in stability (Fig. 5.8).

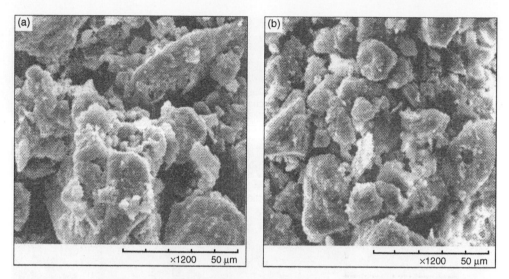

Figure 5.9 SEM images of collapsible loess before and after permeation test: (a) before and (b) after (Luo, 2010).

Many large pores were found in the loess. Generally speaking, the permeability of loess is determined by the number and connectivity of these large pores. In the early stage of the permeability test, water flowed through these large pores, and does not damage them. In this case, the seepage channel is unimpeded and the permeability coefficient is large. However, with the development of permeability, water damages the structure of loess. The fine particles in loess move under the impact of osmotic pressure. These fine particles are not removed from the loess samples with water, but gather in the water channel under the fixation of the instruments and permeable stones. Along with the extension of permeability time, the granular overhead system of loess is destroyed and the effective seepage channel in loess is blocked, resulting in the decrease in the permeability coefficient.

The SEM images magnified 1200 times clearly show the structure changes of the primary loess before and after permeability tests (Fig. 5.9). Before the test, coarse particles account for the vast majority, the pores are large and many and the overall arrangement of the whole structure is loose. After the test, the fine particles fill among coarse particles, large pores are reduced, and the overall structure is more compact than that before the permeability test. These changes above lead to the blockage of water channel in loess, leading to a decrease in permeability coefficient after the test.

Along with the impact of water permeation, the particles collapse or roll down into the pores. The cement between particles softens and dissolves, and chemical changes even occur. Therefore, the cementation strength between particles weakens. The damage of loess structure can further affect the permeability of loess. As shown in Figure 5.10, the black parts, like strips and holes, represent the cracks and pores in the soil, respectively. As the permeation proceeds (from left to right), these fractures and pores gradually decrease and nearly disappear. The very few black hole-like parts

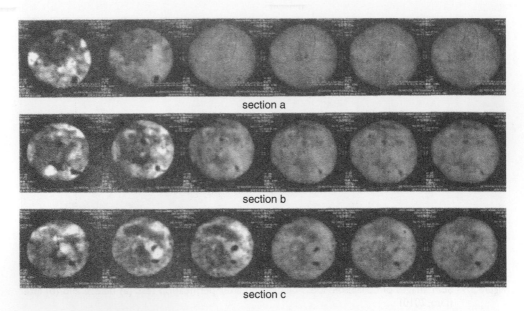

section a

section b

section c

Figure 5.10   CT images of three sections of loess samples at different time during the process of soaking (Fang et al., 2011).

which did not disappear are isolated, closed, and disconnected circular (or elliptic) pores. Such pores have little impact on the permeability. In summary, these structure changes are the essential reasons that the permeability coefficient decreases with the increase of permeation time.

5.5   FREEZE–THAW CYCLE

With the increase of freeze–thaw cycles, the permeability coefficient of loess samples with $\rho_d > 1.3\,\text{g/cm}^3$ and $\omega > 12\%$ increases gradually (Figs. 5.11 & 5.12). The reason is that the volume is enlarged when the water in the loess pores or cracks freeze, which deepens and widens the internal cracks. When the ice melts, the water permeates further inside the soil along the enlarged cracks, and then it is frozen into ice again. The cycle of freezing and melting happens frequently, making the cracks expand continuously and gradually developing a good seepage channel. This phenomenon leads to the increase of loess permeability.

In Figure 5.11, the soil samples with $\rho_d < 1.3\,\text{g/cm}^3$ appear contrary to the law above because the connection between particles is loose and the pores develop for the loess with low dry density. The pores inside the soil have enough space to accommodate the frozen expansion of water. Therefore, the primary structure of loess with such a dry density is not destroyed intensively. With the increase of freeze–thaw cycles, the soil particles aggregate under the cohesion of freezing, the loess samples become consolidated and dense, and the permeability is decreased consequently.

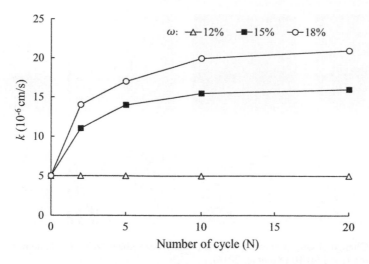

Figure 5.11 Relationship between permeability coefficient and the number of freeze–thaw cycles of samples with different dry densities (Lian et al., 2010).

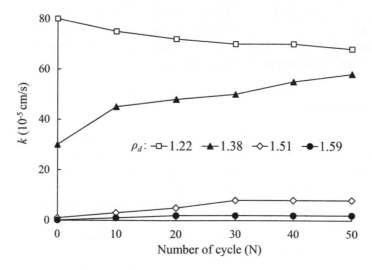

Figure 5.12 Relationship between permeability coefficient and cycle number of samples with different initial water contents (Xu et al., 2016).

As shown in Figure 5.12, the relationship curve between permeability coefficient and cycle number of the sample with $\omega = 12\%$ is a horizontally straight line, indicating that the permeability of this sample does not change. The reason is that the sample moisture content is low, the freezing split produced by ice crystal growth in the pores is weak, and the change of permeability coefficient during freeze–thaw cycles is not significant enough.

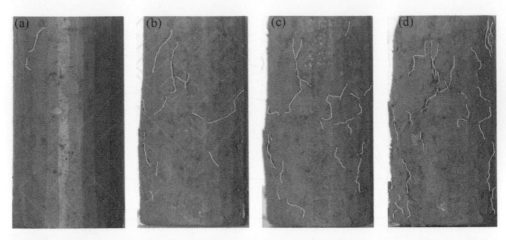

Figure 5.13 Change of saturated loess samples under freeze–thaw cycles: (cycle numbers): (a) 0; (b) 5; (c) 7; and (d) 10 (Xu et al., 2016).

Taking Xi'an Q_3 loess as an example, the structure of loess is damaged seriously by the freeze–thaw cycles. With the increase of freeze–thaw cycles, not only is the surface part of loess samples peeled off, but increasingly irregular cracks also appear on the whole samples (Fig. 5.13). These cracks are good unimpeded seepage channels, and consequently enhance the permeability of loess.

5.6 ANISOTROPY

For primary loess, the permeability coefficient in the vertical direction (k_v) is greater than that in horizontal direction (k_h) (Fig. 5.14). When the dry density is small, the difference between k_v and k_h is large. However, the k_v and k_h tend to be equal as the dry density increases ($\rho_d > 1.60\,\text{g/cm}^3$). For remodeled loess samples, the vertical permeability coefficient (k_v) is smaller than the horizontal permeability coefficient (k_h) (Fig. 5.15), and no obvious regular pattern of difference between k_v and k_h is found.

An obvious difference in the anisotropy of permeability between primary loess samples and remolded loess samples is found. The reason for this difference is closely related to the structure and porosity of samples. When the initial structure of loess is not destroyed, the existence of vertical tubular large pores causes the vertical permeability coefficient to be larger than the horizontal permeability coefficient ($k_v > k_h$). The more large pores are present, the larger the difference. Water mainly flows out of the loess from the horizontal seepage of the remolded loess samples after many repeat compaction and lamination cycles. Therefore, the horizontal permeability coefficient is larger than the vertical permeability coefficient in remolded samples ($k_h > k_v$).

In the SEM images (50 × magnification) of natural loess, many large and round holes (>0.5 mm in diameter) are found in the horizontal layer. By contrast, many large holes are also found in the vertical layer, but with long shapes (≈1 cm in length) (Fig. 5.16). The SEM images (1000 × magnification) show the microstructure

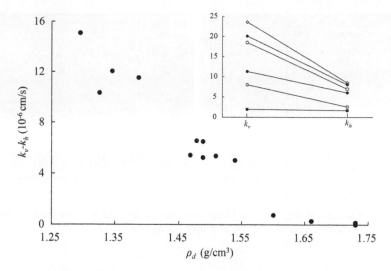

Figure 5.14 Permeability anisotropy of primary loess (Data from: Li, 2007a; Li, 2007b; Liu, 2015).

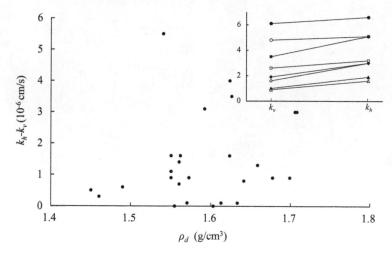

Figure 5.15 Permeability anisotropy of remolded loess (Date from: Guo, 2009; Guo et al., 2009; Guo et al., 2010; Li et al., 2011).

of primary loess (Fig. 5.17). The vertical arrangement of loess particles observed from horizontal slices is loose and large pores exist. The horizontal arrangement of loess particles observed from the vertical slice is relatively tight and the pores are relatively small. The above phenomena show that unimpeded seepage channels in the vertical direction exist when water flows in primary loess, but the horizontal direction lacks such extended interconnecting long holes. This macroscopic and microscopic pore distribution is one of the significant reasons for the anisotropy of loess permeability coefficient.

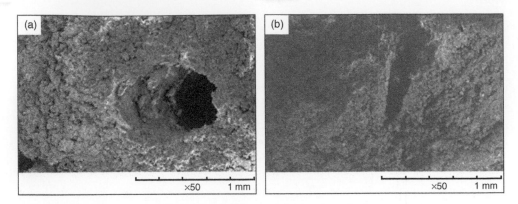

Figure 5.16 SEM images of the horizontal layer and vertical layer of primary loess in Xi'an (50×): (a) horizontal layer; and (b) vertical layer (Liang et al., 2012).

Figure 5.17 SEM images of the horizontal layer and vertical layer of primary loess in Wangjiagou (1000×): (a) horizontal layer; and (b) vertical layer (Wu et al., 2016).

REFERENCES

[1] An, P. (2011) *Experimental study on seepage degradation of reshape saturated loess under seepage.* Msc Thesis. Northwest A&F University, Xianyang.

[2] An, P., Zhang, A.J., Liu, H.T. & Wang, T. (2013) Degradation mechanism of long-term seepage and permeability analysis of remolded saturated loess, *Rock and Soil Mechanics*, (7).

[3] Fang, X.W., Shen, C.N. & Chen, Z.W. (2011) Triaxial wetting tests of intact Q_2 loess by computed tomography (In Chinese). *China Civil Engineering Journal*, 10, 98–106.

[4]   Feng, W., Li, S.S. & Gao, L.X. (2013) Study on relationship between microstructure and soil-water characteristics of remolded clay (In Chinese). *Journal of Guangxi University (Natural Science Edition)*, 01, 170–175.

[5]   Hou, X.K., Li, T.L. & Xie, X. (2016) The effect of undisturbed Q_3 loess's microstructure on its SWCC (In Chinese). *Journal of Hydraulic Engineering*, 47(10), 1307–1314.

[6]   Guo, H. (2009) *Experimental study on triaxial seepage of disturbed Q_3 loess in different regions* (In Chinese). Msc Thesis. Northwest A & F University, Xianyang.

[7]   Guo, H., Luo, Y.S. & Guo, J. (2010) Infiltration characteristics of saturated loess by considering the role of stress field (In Chinese). *Bulletin of Soil and Water Conservation*, 01, 131–133.

[8]   Guo, H., Luo, Y.S. & Li, G.D. (2009) Experimental Research on Triaxial Seepage Test of Saturated Loess Based on Regional Differences (In Chinese). *China Rural Water and Hydropower*, 10, 112–114.

[9]   Li, G.D., Luo, Y.S. & Guo, H. (2011) Study on the triaxial permeability test of Yangling loess (In Chinese). *Yellow River*, 07, 141–143.

[10]   Li, L. (2007a) *Seepage deformation experimental research of fissured loess* (In Chinese). Msc Thesis. Chang'an University, Xi'an.

[11]   Li, P. (2007b) *Research on triaxial seepage test of saturated loess* (In Chinese). Msc Thesis. Northwest A&F University, Xianyang.

[12]   Li, Y.F. (1991) Relationship between the permeability and the porosity of Luochuan's loess layer (In Chinese). *Journal of Earth Sciences and Environment*, 02, 60–64.

[13]   Lian, J.B., Zhang, A.J. & Guo, M.X. (2010) Static and dynamic axial load testing on PHC pipe-piles in deep soft soil (In Chinese). *Yangtze River*, 12, 55–58 & 62.

[14]   Liang, Y., Xing, X.L., Li, T.L., Xu, P. & Liu, S.L. (2012) Study of the anisotropic permeability and mechanism of Q_3 loess (In Chinese). *Rock and Loess Mechanics*, 33(5), 1313–1318.

[15]   Lin, X.Y., Li, T.L., Zhao, J.F., Wang, H. & Li, P. (2014) Permeability characteristics of loess under different consolidation pressures in the Heifangtai platform (In Chinese). *Hydrogeology & Engineering Geology*, 01, 41–47.

[16]   Liu, M.M. (2015) *Hydraulic conductivity variation of loess slope in southern tableland of Jingyang* (In Chinese). Msc Thesis. Chang'an University, Xi'an.

[17]   Luo, J. (2010) Research the characteristics and mechanism of loess seepage deformation (In Chinese). Msc Thesis. China University of Geosciences, Wuhan.

[18]   Wang, H., Yue, Z.R. & Ye, C.L. (2009) Experimental investigation of permeability characteristics of intact and reshaped loess (In Chinese). *Journal of Shijiazhuang Railway Institute (Natural science)*, 02, 20–22 & 31.

[19]   Wang, T., Zhang, A.J., Liu, H.T. & An, P. (2013) Permeability properties of reshaped loess in osmotic solution of different acidities (In Chinese). *Journal of Yangtze River Scientific Research Institute*, 02, 35–40.

[20]   Wang, T.X., Yang, T. & Lu, J. (2016) Influence of dry density and freezing-thawing cycles on anisotropic permeability of loess (In Chinese). *Rock and Soil Mechanics*, (S1), 72–78.

[21]   Wang, Z.J., Chen, Y.P. & Cao, Y. (2007) Triaxial seepage test of saturated loess (In Chinese). *Yellow River*, 12, 87–88.

[22]   Wu, X.G., Liang, Q.G. & Niu, F.J. (2016) Anisotropy of intact loess of Wangjiagou tunnel along Baoji-Lanzhou passenger dedicated line (In Chinese). *Rock and Soil Mechanics*, 08, 2373–2382.

[23]   Xu, J., Wang, Z.Q., Ren, J.W. & Yuan, J. (2016) Experimental research on permeability of undisturbed loess during the freeze-thaw process (In Chinese). *Journal of Hydraulic Engineering*, 47(9), 1208–1217.

[24]   Yang, B., Li, Z. & Dong, X.Q. (2015) Experimental studies on relationship between deformation and hydraulic permeability of loess in process of penetration (In Chinese). *Journal of Guangxi University (Natural Science Edition)*, 02, 325–330.

[25]   Zhang, S.H., Su, N.N. & Dong, X.Q. (2015) Microstructure test of loess under different consolidation pressures (In Chinese). *Science Technology and Engineering*, 15(31), 74–78.

[26]   Zhao, T.Y. & Wang, J.F. (2012) Soil-water characteristic curve for unsaturated loess soil considering density and wetting-drying cycle effects (In Chinese). *Journal of Central South University (Science and Technology)*, 43(6), 2445–2453.

Chapter 6

Shear strength

Shear strength of loess is the strength against the type of yield or failure where the loess fails in shear. It always works as one of the key parameters for evaluation of loess slope stability, design of foundation pit engineering and anti-slide engineering (Liu, 1985; Li, 2010; Cao, 2011). The factors that affect the shear strength of loess can be classified into internal and external factors. The internal factors mainly include the sedimentary age, structure, void ratio, dry density, joint (density and joint surface roughness), and grain composition. The external factors include the moisture content, test methods, shear direction, drainage condition, and temperature (Mi et al., 2006; Li & Miao, 2006; Li, 2004).

The influence of various factors on the shear strength of loess is discussed in the following sections, mainly from the five aspects of loess sedimentary age, moisture content, compaction degree, dry density, void ratio, test methods, and shear direction.

6.1 SEDIMENTARY AGE

Sedimentary age is one of the important factors that affects the shear strength of loess. The older the loess layer, the greater the internal friction angle and cohesion (Figs. 6.3 & 6.4; Table 6.1). This is probably due to the high dry density and low void ratio of older loess sediments (Figs. 6.1 & 6.2).

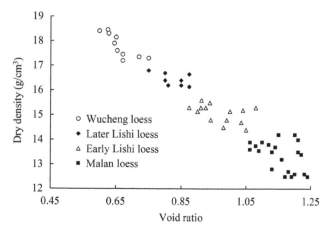

Figure 6.1 Dry density and void ratio of loess with different sedimentary ages (Data from Li et al., 2007).

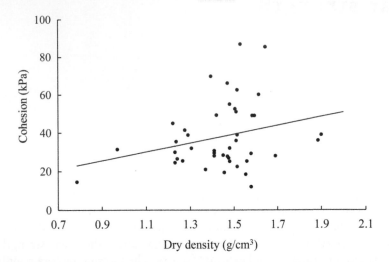

Figure 6.2 Relationship between cohesion and dry density (Data from [3], [8], [10–12], [17], [20], [29], [34], [38], [42], [45], [47–51]).

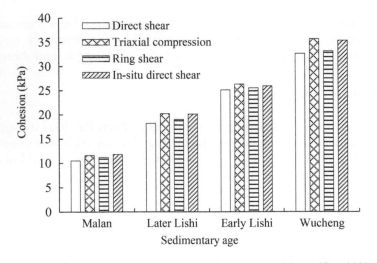

Figure 6.3 Influence of sedimentary age on the cohesion of loess (Cao, 2005).

6.2 MOISTURE CONTENT

Two different views about the influence of moisture content on the shear strength of loess exist. One view is that with the increase of moisture content, the cohesion (c) and internal friction angle (ϕ) decrease, and consequently the shear strength of loess decreases (Mi et al., 2006; Dang & Li, 2001; Wu, 2007). The other view is that the cohesion (c) increases first and then decreases with the increase of moisture content, and the changing trend of internal friction angle is just opposite to cohesion

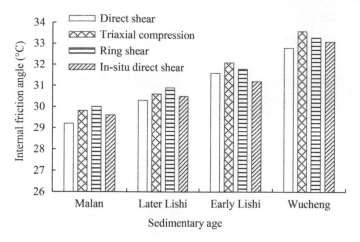

Figure 6.4 Influence of sedimentary age on the internal friction angle of loess (Cao, 2005).

Table 6.1 Range of cohesion and internal friction angles of loess with different sedimentary ages (Statistic analysis on data from [3], [9], [11–12], [14], [20], [29–32], [34], [38], [42], [47–48] and [51]).

	Cohesion (kPa)				Internal friction angle (°)			
Stratum	Maximum value	Minimum value	Concentration range	Average	Maximum value	Minimum value	Concentration range	Average
Q_4	63	0.686	15–25	21.09	30.7	10.2	17–20	18.42
Q_3	92.68	9	30–40	34.82	37.2	14.3	23–25	23.96
Q_2	110	25.2	35–45	48.18	50	18.5	30–33	29.77
Q_1	393.2	207.4	250–350	296.8	40	26.3	30–33	31.76

(Li & Miao, 2006; Li et al., 2007; Xu, 2013). The shear strength depends on both the c and ϕ of loess.

The scholars with the first view think that the influence of the moisture content on shear strength of loess can be explained from the following two aspects. The first is that with the increase of moisture content, the bonding and van der Waals forces between particles decrease because the soluble component in loess is easily dissolved in water. The second is that the thickness of the bound water film on the surface of particles increases with the increase of water molecules in loess pores, which makes the strongly bound water gradually transform into weakly bound water, and the weakly bound water into free water. These two aspects lead to the decrease of cohesion and internal friction angle of loess, and consequently the decrease of loess shear strength (Figs. 6.5 & 6.6).

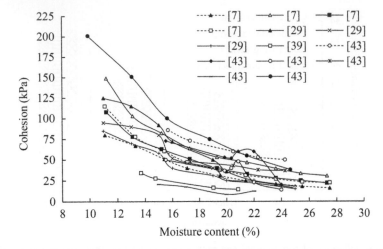

Figure 6.5 Relationship between cohesion and moisture content of loess (Data from [7], [29], [39], [43]).

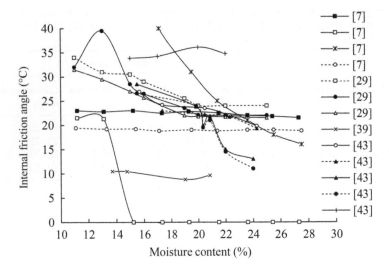

Figure 6.6 Relationship between internal friction angle and moisture content of loess (Data from [7], [29], [39], [43]).

However, the scholars with the other view think that the loess particles are mainly cemented by clay minerals, free oxides, carbonate, and organic matter, and the cohesion and internal friction angle of loess mainly depend on the strength of cements between particles. When the moisture content of loess is less than a certain value (approximately 10%), the cementation increases gradually with increase of the moisture content, resulting in the increase of cohesion and internal friction angle of loess. When the moisture content of loess is greater than a certain value, the cementation

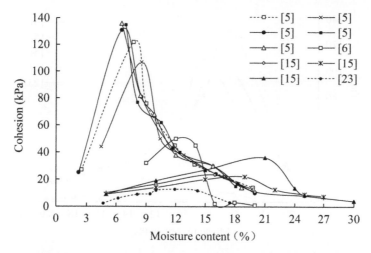

Figure 6.7 Relationship between cohesion and moisture content of loess (Data from [5–6], [15], [23]).

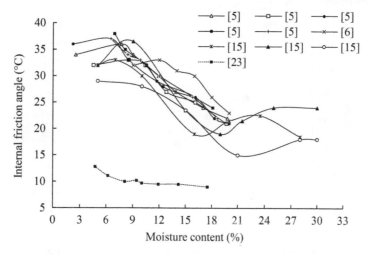

Figure 6.8 Relationship between internal friction angle and moisture content of loess (Data from [5–6], [15], [23]).

decreases and the c, ϕ values gradually decrease with the increase of water content, because part of the cementing materials between particles will be gradually dissolved in water. Therefore, the c and ϕ of loess increase first and then decrease with the increase of moisture content. Moreover, the c, ϕ of loess will gradually tend to be stable when the moisture content is more than a certain value (approximately 20%) (Figs. 6.7 & 6.8).

The reason for differences in the two views may be the different initial moisture content of loess samples they used. The scholars with the first view used samples with large moisture content, ignoring the relationship between the moisture content and c, ϕ of loess with small moisture content. The latter view is based on samples with a

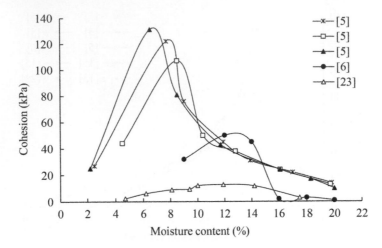

Figure 6.9 Relationship between cohesion and moisture content in the direct shear test (Data from [5–6], [23]).

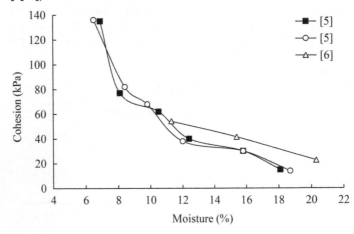

Figure 6.10 Relationship between cohesion and moisture content in the triaxial compression test (Data from [5–6]).

wide range of moisture content, showing a comprehensive relationship between the moisture content and shear strength of loess.

6.3 TEST METHODS

Direct shear, triaxial compression, ring shear, and in situ large-area direct shear tests are often used for measuring shear strength of loess. However, the test results show great differences when different test methods are employed (Figs. 6.3 & 6.4).

In the direct shear test, the cohesion increases first, reaches the peak value, and then gradually decreases with the increase of moisture content (Fig. 6.9). In the triaxial compression, the cohesion decreases with the increase of moisture content (Fig. 6.10).

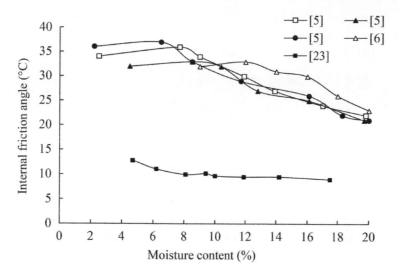

Figure 6.11 Relationship between the internal friction angle and moisture content of loess in the direct shear test (Data from [5–6], [23]).

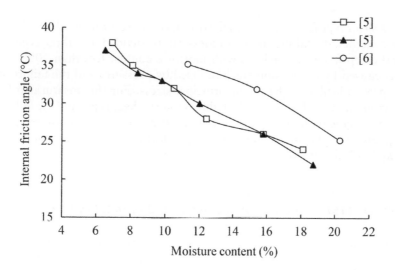

Figure 6.12 Relationship between the internal friction angle and water content of loess in the triaxial shear test (Data from [5–6]).

In direct shear, the moisture content of samples has little influence on the internal friction angle (Fig. 6.11), but has a larger influence on the internal friction angle in triaxial compression (Fig. 6.12). However, the overall trend of the two is constant, and the internal friction angle gradually decreasing with the increase of moisture content.

Different test methods also lead to different shear failure modes of loess. In direct shear test, the loess samples are destroyed along a fixed shear surface, and the loess

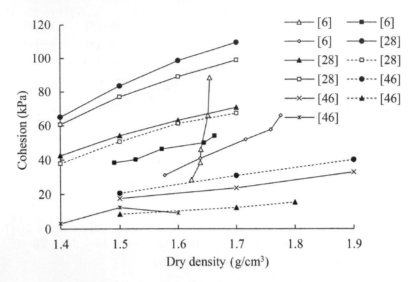

Figure 6.13 Relationship between cohesion and dry density of loess (Data from [6], [28], [46]).

possesses a shear type failure mode. In the triaxial compression test affected by different moisture contents, the failure modes of loess can be divided into three types: fracture, shear, and creep failures. For loess with moisture content less than 10%, the failure of loess is caused by the tension crack induced by pressure, and the failure surface is basically perpendicular to the minor principal stress. For the unsaturated loess with moisture content more than 10%, the failure mode is shear type, and the failure surface is an inclined plane with an angle greater than 45degrees with minor principal stress. For saturated loess, the failure model is creep deformation, and the middle part of the loess sample is bulging. No obvious fracture surface exists.

6.4 DRY DENSITY, VOID RATIO AND COMPACTION DEGREE

Cohesion and internal friction angle of loess increase with the increase of compaction degree and dry density and decrease of void ratio (Wang et al., 2014a; Zhang et al., 2014; Jia et al., 2014; Cheng, 2009; Xu, 2013). The details are presented in the following.

The cohesion and internal friction angle increase with the increase of loess dry density (Figs. 6.13 & 6.14). The greater the dry density, the closer the contact of loess particles, and the smaller the void ratio. This phenomenon is conductive for loess water surface tension to play a role, consequently increasing the cohesion. In addition, the interlocking effect, gravitation between particles, as well as the shear strength also increase.

The cohesion and internal friction angle of loess increase with the increase of compaction degree (Figs. 6.15 & 6.16). Under the same moisture content, the dry

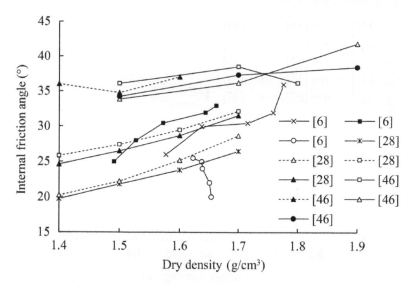

Figure 6.14 Relationship between internal friction angle and dry density of loess (Data from [6], [28], [46]).

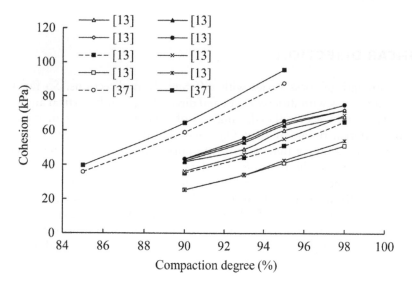

Figure 6.15 Relationship between cohesion and compaction degree of loess (Data from [13], [37]).

density of loess increases with the increase of compaction degree. The contact between particles become closer and the void ratio decreases. The water in the loess is mainly in the form of strongly bound water which cannot move around particles, and the water film is thin. This leads to the reduction of lubrication effect between particles, resulting in the increase of cohesion and internal friction angle.

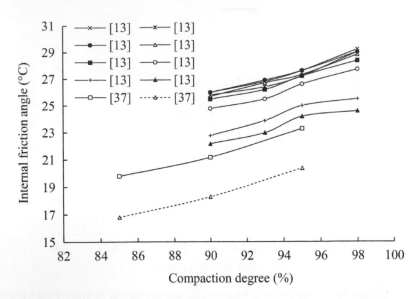

Figure 6.16 Relationship between internal friction angle and compaction degree of loess (Data from [13], [37]).

6.5 SHEAR DIRECTION

The shear strength of loess varies with different shear directions. So far, the shear strength data of loess with different ages and directions (mainly vertical and horizontal direction) have been accumulated. Analysis of this large number of data provides two totally different points of view. One view is that the loess shear strength of the vertical direction is larger than that of the horizontal direction. The other view is completely opposite to the first one. Scholars with the first view believe that in the vertical direction, the connection and the interlocking force between loess particles are weak because of the horizontal deposition of loess (Xu et al., 2015; Zhang et al., 2015; Chen et al., 2015; Yang, 2015). While the scholars with the second view think that the cohesion between vertical particles is larger than that between the horizontal directions because the joints and pores of loess are more developed in the vertical direction than in the horizontal direction (Liu et al., 2002; Xu, 2012; Shao et al., 2014; Ye et al., 2014; Liang et al., 2012a).

6.5.1 Wucheng and Early Lishi loess

For the Wucheng loess and Early Lishi loess, both horizontal and vertical shear strength increase with the normal stress (Fig. 6.17). The shear strength of the vertical and horizontal directions is almost equal for Early Lishi loess (Fig. 6.18). For the Wucheng loess, the shear strength of the vertical direction is slightly larger than that of the horizontal direction. Furthermore, with the increase of normal stress, the shear

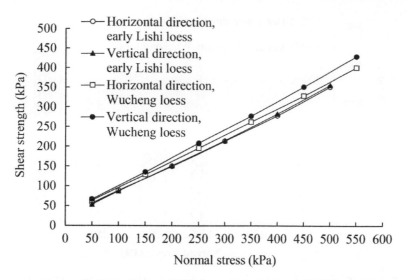

Figure 6.17 Relationship between shear strength and shear direction of Wucheng and Early Lishi loess (Cao, 2005).

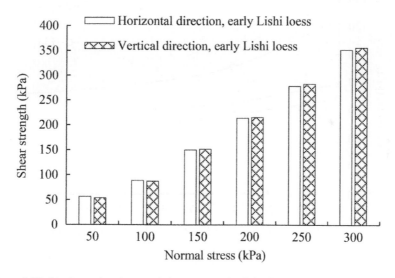

Figure 6.18 Horizontal and vertical shear strength of the Early Lishi loess (Cao, 2005).

strength and the difference between the horizontal and vertical shear strength increase gradually (Fig. 6.19). For the Wucheng and Early Lishi loess with void ratio <1 and dry density >1.5 g/cm³, both cohesion and internal friction angle decrease gradually with the increase of angle between shear direction and vertical direction (Figs. 6.20 & 6.21), showing that the shear strength of vertical direction is larger than that of the horizontal direction.

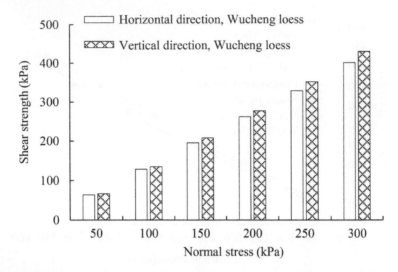

Figure 6.19 Horizontal and vertical shear strength of Wucheng loess (Cao, 2005).

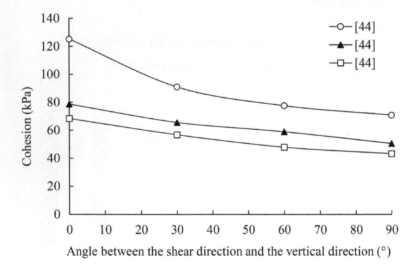

Figure 6.20 Relationship between cohesion and shear direction of loess (Data from [44]).

6.5.2   Later Lishi and Malan loess

For the Malan loess and Later Lishi loess, both horizontal and vertical shear strength increase with the normal stress (Fig. 6.22). The shear strength of vertical direction is less than that of horizontal direction for Malan loess. Moreover, both the shear strength and the difference between the horizontal and vertical shear strength increase gradually with the increase of normal stress (Figs. 6.23 & 6.24). For the Malan and Later Lishi loess with void ratio >1 and dry density <1.5 g/cm³, the cohesion and

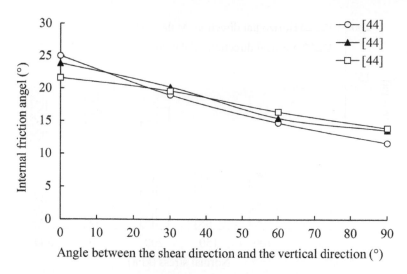

Figure 6.21 Relationship between internal friction angle and shear direction of loess (Data from [44]).

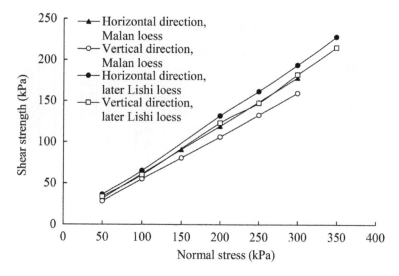

Figure 6.22 Relationship between shear strength and shear direction of Malan and Later Lishi loess (Cao, 2005).

internal friction angle decrease gradually with increase of the angle between the shear direction and horizontal direction (Figs. 6.25 & 6.26). This phenomenon shows that the horizontal shear strength is larger than the vertical shear strength.

As shown in Figure 6.27, with the sequence of the sedimentary age (Lishi loess and Wucheng loess), the permeability coefficient of loess is lower, and the horizontal permeability coefficient is larger. While for Malan loess, the vertical permeability is larger than the horizontal permeability.

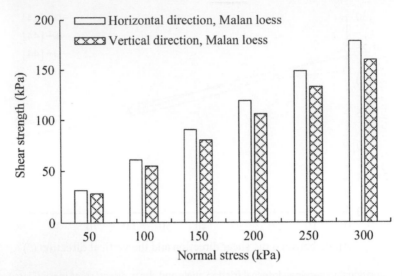

Figure 6.23 Horizontal and vertical shear strength of Malan loess (Cao, 2005).

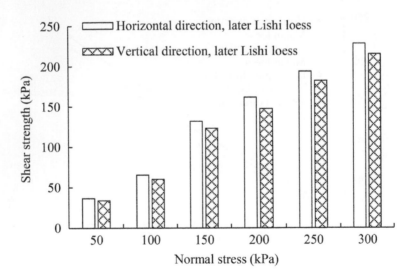

Figure 6.24 Horizontal and vertical shear strength of Later Lishi loess (Cao, 2005).

Figures 6.28 and 6.29 show the SEM images of the microstructure characteristics of loess. The shape of visible large pores in horizontal plane are basically circular (diameter <0.5 mm). In the vertical plane, the circular holes are less and most images have visible long holes. The length of most pores is less than 2 cm, especially close to 1 cm. The direction is vertical or approximately 45 degrees with the vertical direction, indicating that the large pores in the vertical direction are developed. According to statistics, the average number of circular holes in the horizontal plane

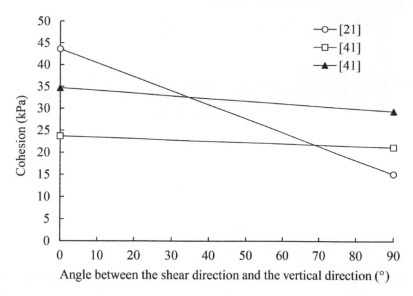

Figure 6.25 Relationship between cohesion and shear direction of Malan and Later Lishi loess (Data from [21], [44]).

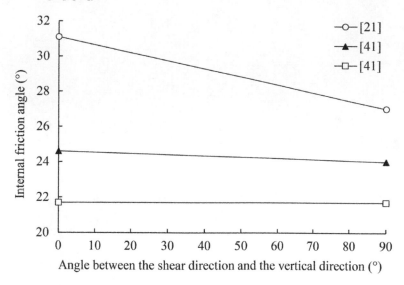

Figure 6.26 Relationship between the internal friction angle and shear direction of Malan and Later Lishi loess (Data from [21], [44]).

is 0.3277 nos./cm², approximately 10 times that of the vertical plane, which has an average number of 0.0384 nos./cm² (Liang et al., 2012b).

All the test results shown in this section (Figs. 6.17 to 6.29) indicate that the shear strength of loess has obvious anisotropy in different directions because of the directional anisotropy of loess structure. The anisotropy of shear strength of loess

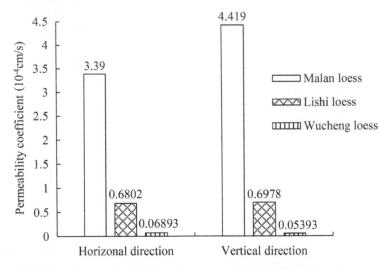

Figure 6.27 Permeability of loess in different directions and with different ages (Cao, 2005).

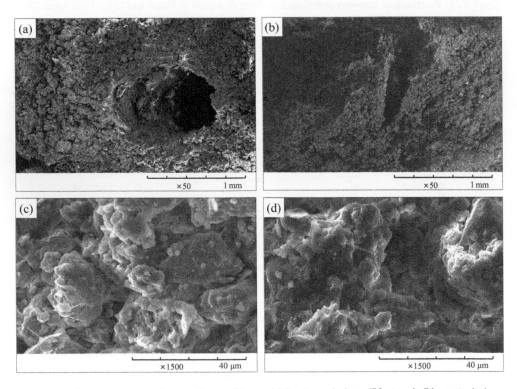

Figure 6.28 SEM images in different planes of loess: (a) horizontal plane (50 times); (b) vertical plane (50 times); (c) horizontal plane (1500 times); and (d) vertical plane (1500 times) (Liang et al., 2012b).

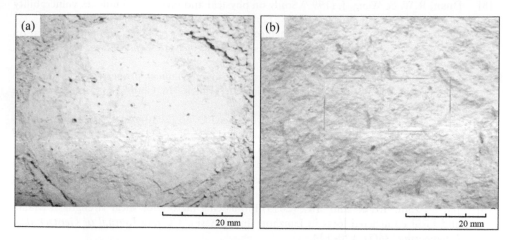

Figure 6.29 Digital images in different planes of loess: (a) horizontal plane; and (b) vertical plane (Liang et al., 2012b).

is not obvious when the moisture content is high (>14%). However, for primary loess, the anisotropy would be significant with natural moisture contents or when the moisture content is low. Generally, the horizontal shear strength is larger than the vertical shear strength for the Malan loess and Later Lishi loess, and it is opposite for the Wucheng and Early Lishi loess. This difference is probably because the vertical joints and vertical pores of Malan loess. The vertical joints of Wucheng and Early Lishi loess are not developed as the Malan loess and Later Lishi loess due to it being in a state of overconsolidation under the long-term effect of the overlying crustal stress (Cao, 2005).

REFERENCES

[1]   Cao, X.P. (2005) Study on mechanical properties of loess in Lanzhou area and its application (In Chinese). Msc Thesis. Lanzhou University, Lanzhou.
[2]   Cao, X.Y. (2011) *Influence of water content and stress on the shear strength of the Loess in the west of Shaanxi Province* (In Chinese). Msc Thesis. University of Chang'an, Xi'an.
[3]   Chen, H. (2011) Multi factor experimental study on static characteristics of saturated loess (In Chinese). *Railway Survey and Design*, (5), 45–49.
[4]   Chen, W., Zhang, W.Y., Chang, L.J., Jie, Y.X., Ma, Y.X. & Wang, M. (2015) Experimental study of anisotropy of compacted loess under directional shear stress path (In Chinese). *Chinese Journal of Rock Mechanics and Engineering*, 34(2), 4320–4324.
[5]   Cheng, B. & Lu, J. (2009) Experimental study on the effect of water content on shear strength of Q_3 loess in North Shaanxi Province (In Chinese). *Construction Technology*, 38, 40–43.
[6]   Cheng, X.Y. (2009) *Experimental study of effects of moisture content on loess strength* (In Chinese). Msc Thesis. China University of Geosciences, Wuhan.
[7]   Dang, J.Q. & Li, J. (2001) Structure strength and shear strength of unsaturated loess (In Chinese). *Journal of Hydraulic Engineering*, 7, 79–84.

[8] Duan, R.W. & Wang, J. (1997) Study on physical and mechanical indexes vulnerability analysis of loess (In Chinese). *China Earthquake Engineering Journal*, (3), 81–85.

[9] Duan, Z.F. & Xu, H.A. (1992) Engineering geological characteristics and distribution law of soft loess in Xi'an area (In Chinese). *Journal of Earth Sciences and Environment*, (1), 64–70.

[10] Feng, M.Y. (2013) Study on the physical and mechanical properties of collapsible loess in the Chengchi high-speed Weichang branch (In Chinese). *Communications Standardization*, (20), 37–39.

[11] Hua, Z.M., Zhang S.A., Zhang, E.X. & Shen, Q.W. (2010) Analysis of the foundation design of high-rise buildings in Longdong Loess Plateau (In Chinese). *Geotechnical Investigation & Surveying*, S1, 267–277.

[12] Jia, H., He, Y.L. & Yang, S.X. (2011) Effects of joints on the damage morphology and mechanical properties of loess (In Chinese). *China Rural Water and Hydropower*, (11), 77–81.

[13] Jia, L., Zhu, Y.P. & Zhu, Y.C. (2014) Influencing factors for shear strength of Malan and Lishi compacted loess in Lanzhou (In Chinese). *Chinese Journal of Geotechnical Engineering*, 36(2), 120–124.

[14] Li, B.X. & Li, Y.J. (2003) Engineering geological characteristics of Malan loess in Lanzhou (In Chinese). *Journal of Gansu Sciences*, 15(3), 31–34.

[15] Li, B.X. & Miao, T.D. (2006) Research on water sensitivity of loess shear strength (In Chinese). *Chinese Journal of Rock Mechanics and Engineering*, 25(5), 1003–1008.

[16] Li, B.X., Niu, Y.H. & Miao, T.D. (2007) Water sensitivity of Malan loess in Lanzhou (In Chinese). *Chinese Journal of Geotechnical Engineering*, 29(2), 294–297.

[17] Li, G.W. & Zhao, J. (2008) Physical and mechanical properties and index analysis of wet and soft foundation in gully region of Loess Plateau (In Chinese). *Shanxi Architecture*, 34(34), 99–100.

[18] Li, G.X. (2004) *Advanced Loess Mechanics* (In Chinese). Beijing, Tsinghua University Press.

[19] Li, W. (2010) *Moisture content is the influence of the shear strength of Gansu loess slope research* (In Chinese). Msc Thesis. Chang'an University, Xi'an.

[20] Li, Y.L. (1995) Study on the basic characteristics and engineering geological properties of loess in the western Henan (In Chinese). *Journal of North China Institute of Water Conservancy and Hydroelectric Power*, (2), 31–37.

[21] Liang, Q.G., Zhao, L., An, Y.F. & Zhang, Y.J. (2012a) Preliminary study of anisotropy of Q_4 loess in Lanzhou (In Chinese). *Rock and Loess Mechanics*, 30(1), 90–96.

[22] Liang, Y., Xing, X.L., Li, T.L., Xu, P. & Liu, S.L. (2012b) Study of the anisotropic permeability and mechanism of Q_3 loess (In Chinese). *Rock and Loess Mechanics*, 33(5), 1313–1318.

[23] Liao, H.J., Li, T. & Peng, J.B. (2011) Study of strength characteristics of high and steep slope landslide mass loess (In Chinese). *Rock and Loess Mechanics*, 32(7), 1939–1944.

[24] Liu, F.Y., Ji, L.C.J., Che, A.L. & Yan, J.W. (2002) Preliminary experimental study on anisotropic properties of Q_3 loess (In Chinese). In: Proceedings of the Seventh Symposium on rock mechanics and Engineering Society of China, *10–12 September 2002, Xi'an*. Beijing, Science and Technology of China Press. pp. 129–131.

[25] Liu, T.S. (1985) *Loess and Environment* (In Chinese). Beijing, Science Press.

[26] Mi, H.Z., Li, R.M. & Niu, J.X. (2006) The influence of water content on shear strength characteristics of Lanzhou intact loess (In Chinese). *Journal of Gansu Sciences*, 18(1), 78–81.

[27] Shao, S.J., Xu P., Wang, Q. & Dai, Y.F. (2014) Study on the loess anisotropy by true triaxial test (In Chinese). *Chinese Journal of Geotechnical Engineering*, 36(9), 1614–1623.

[28] Wang, J.J., Zhang, X.L. & Wang, T.X. (2014a) Study on the shear strength characteristics of compacted loess considering the influence of water content and density (In Chinese). *Xi'an University of Architecture & Technology (Natural Science Edition)*, 46(5), 687–691.

[29] Wang, X.J., Mu, N.S. & Wu, Q. (2014b) Statistical characteristics of physico-mechanical indices of the new loess (In Chinese). *Highway Engineering*, (1), 88–93.

[30] Wang, Z. (2011) Experimental study on consolidation of collapsible loess subgrades by impact rolling compaction on Shijiazhuang-Taiyuan passenger-dedicated line (In Chinese). *Railway Standard Design*, (9), 30–32.

[31] Wei, J.W. (2009) Discussion on the change of mechanical parameters of collapsible loess under humidification condition (In Chinese). *Site Investigation Science and Technology*, (5), 41–44.

[32] Wu, Y.P. & Zhao, C.X. (2001) The engineering geological properties of Fulongping District Loess in Lanzhou City (In Chinese). *Journal of Gansu Sciences*, 13(4), 41–43.

[33] Wu, Z.G. (2007) *Experimental study on structural strength and shear strength of the unsaturated loess* (In Chinese). Msc Thesis. Northwest A&F University, Yangling.

[34] Xie, F.C. & Jiang, Z.F. (1987) Basic characteristics of loess in West Henan (In Chinese). *Henan Geology*, (3), 44–50.

[35] Xu, P. (2012) *Study on the loess anisotropy by true triaxial test* (In Chinese).Msc Thesis. Xi'an University of Science and Technology, Xi'an.

[36] Xu, S.C., Liang, Q.G., Li, S.S., Zhang, T.J. & Zhang, R. (2015) Experimental study on anisotropy of undisturbed Q_3 loess in Dingxi, Gansu (In Chinese). *Journal of Geomechanics*, 21(3), 378–385.

[37] Xu, Y.Y. (2013) *The analysis of temperature effect and experimental research on collapsible loess of Wangpogou Bridge* (In Chinese). Msc Thesis. Chongqing University, Chongqing.

[38] Yang, F., Chang, W., Wang, F.W. & Li, T.L. (2014) Numerical simulation of Dongfeng high speed remote loess landslide movement process in Shaanxi Jingyang (In Chinese). *Journal of Engineering Geology*, 22(5), 890–895.

[39] Yang, L. (2010) Analysis of Characteristics about Shear Strength of Loess (In Chinese). *Journal of Water Resources and Architectural Engineering*, 8(3), 163–166.

[40] Yang, Z.X. (2015) *Experimental study of strength and deformation anisotropy of compacted loess* (In Chinese). Msc Thesis. Xi'an University of Science and Technology of Architecture, Xi'an.

[41] Ye, C.L., Zhu, Y.Q., Liu, Y.J. & Song, Y.X. (2014) Experimental study on the anisotropy and unloading deformation characteristics of intact loess (In Chinese). *China railway science*, 35(6), 1–6.

[42] Yu, Y.H. & Yang, X.H. (2003) Experimental study on mechanical properties of cement loess (In Chinese). *Journal of Chang'an University (Natural Science Edition)*, 23(6), 29–32.

[43] Zhang, B.P., Wang, L. & Yuan, H.Z. (1994) The Influence of Moisture Content on Stucture Strength Characteristics of Loess (In Chinese). *Journal of Northwest Science Technology University of Agriculture and Forestry (Natural Science Edition)*, 22(1), 54–59.

[44] Zhang, B.Q., Lou, Z.K., Ma, X. & Zhang, N.N. (2015) Experimental study on anisotropy characteristics of loess under plane strain condition (In Chinese). *Yangtze River*, 46(7), 55–59.

[45] Zhang, L.J., Mi, H.Z. & Li, J.X. (2004) Statistical analysis of mechanical properties of loess in the third terrace of Lanzhou (In Chinese). *Journal of Gansu Sciences*, 16(1), 113–115.

[46] Zhang, P.R. Huang, Z. Yang, F. & Xiao, D. (2014) Study of reshape loess shear strength (In Chinese). *Shanxi Architecture*, 40(20), 71–73.

[47] Zhang, S.A. (1994) Engineering characteristics and nonlinear parameters determination of the old loess in Jiuzhoutai section (In Chinese). Hydrogeology and Engineering Geology, (5), 22–25.

[48] Zhang, Z.L. (2003) The engineering characteristics of collapsible loess in Tianshui (In Chinese). *Geotechnical Engineering World*, 6(9), 71–73.

[49] Zhang, Z.Q., Jing, J. & Fu, H.B. (2007) Experimental study on physical and mechanical properties of collapsible loess subgrade (In Chinese). *Journal of Highway and Transportation Research and Development*, 24(7), 48–51.

[50] Zhi, J.C. (2011) A preliminary study on the influences of the collapsibility coefficient and immersion on mechanical properties of loess (In Chinese). *Shanxi Hydrotechnics*, (1), 9–10.

[51] Zhou, Q., Zhao, F.Z. & Zhang, H.L. (2006) Effects of compaction degree and water content on mechanical properties of compacted loess (In Chinese). *Highway*, (1), 67–70.

Chapter 7

Tensile strength

7.1 SIGNIFICANCE

Previous studies on the mechanical properties of loess often focus on the shear strength. Research on tensile strength, by contrast, is very limited. Although the tensile strength of loess is small, it is closely related to the development and occurrence of geohazards, such as landslides, fallings, and ground fissures.

In general, intermittent tensile cracks tend to form on the rear parts of the slope when the tensile stress of the trailing edge exceeds the tensile strength of slope soil. This, in return, provides conditions for further occurrence of collapse or landslide. Tensile failures may occur in the core of a high earth-rock dam due to the arc effect produced by the settlement of soil. Additionally, the foundation is often damaged by tensile failures because of the difference of load and the uneven ground.

China has the largest accumulation in loess distribution throughout the world. As the main material of mountain slopes, building foundations, road foundations, and dam bodies in loess areas, the mechanical properties of loess are very important to engineering construction and human life. Given the frequent occurrence of geological disasters in loess areas and the frequent tensile failure phenomenon of loess foundations and slopes, a detailed study on the tensile properties and the mechanism of tensile failure of loess can provide reference and guidance to the design and construction of engineering projects.

7.2 TENSILE STRENGTH

7.2.1 Cohesive

Loess cohesive force can be divided into the original cohesion and the solidified cohesion according to the formation mechanism and formation time of loess.

The cohesive force of remolded loess belongs to the original cohesion, which is caused by the effect of orientation, induction, and dispersion of charged particles in loess. The factors that mainly affect the original cohesion are the property (whether they are charged and the number of charges) and the space of particles.

Solidified cohesion is the increased strength in the process of soil formation under the condition of constant dry density and water content. Its numerical value is the difference in cohesive force between undisturbed loess and remolded loess. The formation causes of the solidified cohesion include primary cementation, recrystallization, and subsequent cementation. The primary cementation includes the precipitation of iron and silicon oxide, the precipitation of slightly soluble and soluble salt, and cementation formed by biochemical action. Recrystallization is the process of crystallization of fine-grained clay minerals and secondary oxides, forming clay minerals into aggregates. The subsequent cementing materials are the deposition of calcite, secondary oxide colloids, and soluble salts under the action of water.

Greater the dry density of loess indicates greater original cohesion. The solidified cohesion formed by various cementing materials at the contact point between particles tends to decrease gradually when the natural structure of loess is destroyed under the action of external forces. The cohesion of unsaturated loess decreases with the increase of moisture content, and it almost completely disappears when the loess is saturated.

7.2.2 Values of tension strength

The tensile strength of loess is the ultimate resistance to external tensile force, thus reflecting the magnitude of loess cohesion. The tensile strength and ultimate tensile strain of loess are small. In uniaxial tension test, the tensile strength of undisturbed loess is generally in the range of 0 kPa to 72.2 kPa, and the tensile strength of remolded loess is in the range of 0 kPa to 23 kPa. The test result based on the three-point bending method is larger than that by other methods. Three-point bending shows that the tensile strength of undisturbed loess ranges from 0 kPa to 221 kPa, and the tensile strength of remolded loess ranges from 0 kPa to 161 kPa.

7.3 INFLUENCING FACTORS

Studies show that the main factors affecting the tensile strength of loess are the moisture content and dry density. In addition, the testing methods of tensile strength, the size of loess specimen, and the stretching (fracturing) rate all affect the test results.

7.3.1 Moisture content

Generally, the tensile strength decreases with the increase of moisture content when the dry density of loess is constant. The relation between tensile strength and moisture content is approximately a negative exponential (Sun et al., 2009a, 2009b; Sun et al., 2010; Yuan et al., 2015; Wang et al., 2015).

The impact of moisture content on the strength of loess body is mainly attributed to the form of water in loess, the interaction between water and soil, the contact state between particles, and the changes of loess structure. When the water content of loess is greater than its liquid limit, the soil particles are completely separated by the free water, and the soil body has fluidity, leading to disappearance of tensile strength. The soil body is in a state of plasticity when the water content is between the liquid limit and plastic limit. The soil particles are mainly connected by weak-binding-water, and its connection strength is relatively weak. With the decrease of moisture content, the thickness of weak-bound-water film decreases, leading to the gradual increase of the force between particles and tensile strength of loess. The soil water mainly becomes strong-binding-water when the moisture content further decreases and reaches the

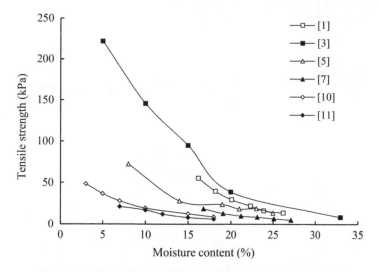

Figure 7.1 Relationship between the tensile strength of undisturbed loess and moisture content (Data from [1], [3], [5], [7], [10–11]).

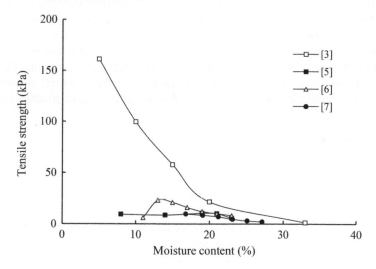

Figure 7.2 Relationship between the tensile strength of remolded loess and moisture content (Data from [3], [5–7]).

plastic limit. The contact area and force between particles increase because of the formation of soil particle aggregates, further leading to tensile strength increase.

As shown in Figures 7.1 and 7.2, the changing trend of the tensile strength of undisturbed loess and remodeled loess is almost consistent with moisture content. That is, the tensile strength decreases with the increase of moisture content, and eventually tends to approach 0 kPa.

For undisturbed loess, the tensile strength is in the range of 0 kPa to 221 kPa when the moisture content is 3% to 33%. While for remodeled loess, the tensile strength

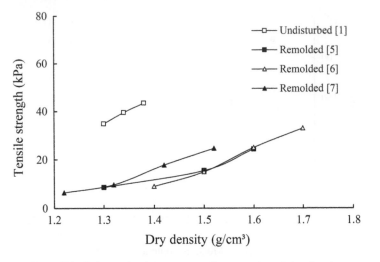

Figure 7.3 Relationship between tensile strength and dry density.

falls into the range of 0 kPa to 161 kPa when the moisture content is 5% to 33%. By comparison, the tensile strength of the remolded loess is significantly lesser than that of the undisturbed loess, and the changes of moisture content affects the tensile strength of undisturbed loess. The slope of the curve is large at a low level of moisture content, and it gradually decreases with the increasing water content. The curve of the remolded loess is relatively smooth, indicating that the moisture content has less influence on the tensile strength of remodeled loess than the undisturbed loess.

7.3.2 Dry density

When the moisture content is constant, the tensile strength of loess increases with the increase of dry density (Sun et al., 2009a, 2009b; Luo & Xing, 1998; Sun et al., 2006).

The space between soil particles is reduced and the soil arrangement becomes closer with the increase of dry density. The original cohesion will be increased by the interaction of adhesion force and intermolecular force between mineral particles. In addition, the adsorptive intensity formed by matrix suction also increases with the increase of loess dry density when moisture content is constant.

Figure 7.3 shows that the tensile strength of loess increases with dry density, and the relationship is approximately linear. The tensile strength of remolded loess is in the range of 6.4 kPa to 33 kPa when the dry density of 1.22 g/cm^3 to 1.70 g/cm^3. For the undisturbed loess, the tensile strength falls into the range of 34.9 kPa to 43.5 kPa when dry density ranges from 1.30 g/cm^3 to 1.38 g/cm^3. The tensile strength of the undisturbed loess is obviously higher than that of remolded loess.

7.3.3 Specimen size

The tensile strength gradually decreases and finally stabilizes with the increase of length-to-diameter ratio of the specimen. However, the influence of specimen diameter

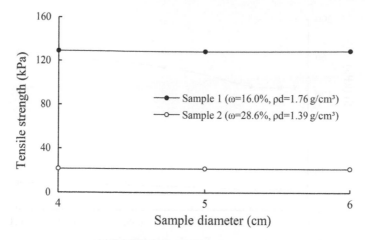

Figure 7.4 Relationship between tensile strength and sample diameter (Data from: Lu et al., 2015).

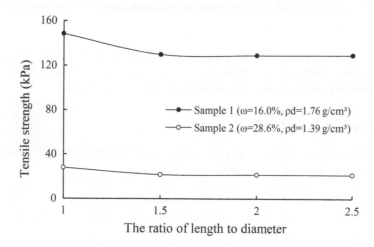

Figure 7.5 Relationship between tensile strength and the ratio of length to diameter (Lu et al., 2015).

on tensile strength is not obvious. The specimen with large diameter is recommended for a facilitative operation of experiments (Lu et al., 2015).

Figure 7.4 shows that the tensile strength of loess is nearly constant when the sample length is fixed and the diameter is in the range of 4 cm to 6 cm. Figure 7.5 shows that the tensile strength decreases with the increase of the ratio of length to diameter when the ratio varies between 1 and 1.5 and the diameter is constant. When the ratio of length to diameter is larger than 1.5, the curve is almost horizontal and the tensile strength tends to be stable.

In addition, when the sample length is constant and the diameter is in the range of 4 cm to 6 cm, the tensile strength of sample 1 is 129.0 kPa and 21.5 kPa for sample 2. When the diameter of samples is constant and the ratio of length to diameter is in range of 1 to 2.5, the tensile strength of sample 1 is 129.0 kPa to 148.3 kPa and 21.4 kPa

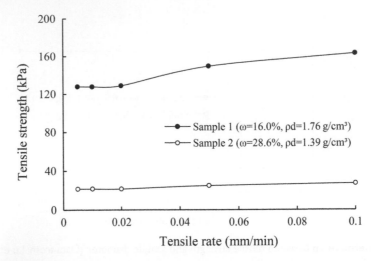

Figure 7.6 Relationship between tensile strength and tensile rate (Lu et al., 2015).

to 21.8 kPa for sample 2, and the slope of curve 1 is larger than that of curve 2. The difference in test results of tensile strength in sample 1 and sample 2 results from the different moisture contents and dry densities of the two samples.

7.3.4 Tension speed

The uniaxial tensile strength of loess increases and the ultimate tensile strain decreases with the increase of tensile rate (Lu et al., 2015).

The tensile strength of loess is influenced by the internal pore water pressure. The pore water pressure in loess dissipates gradually under a small tensile rate, while it cannot dissipate in time under a larger tensile rate, leading to the greater tensile strength at large tensile rates than that at small tensile rates.

Figure 7.6 shows that the tensile strength of sample 1 is 127.9 kPa to 162.8 kPa when the tensile rate is in the range of 0.005 mm/min to 0.1 mm/min, and the tensile strength of sample 2 is 21.4 kPa to 27.2 kPa under the same condition. The tensile strength does not change when tensile rate is less than 0.02 mm/min, and its curve is close to being horizontal. When the tensile rate exceeds 0.02 mm/min, the curve shows an upward trend and the tensile strength of loess increases with tensile rates. The slope of curve 1 is larger than that of curve 2.

7.3.5 Consolidation pressure and compaction degree

The tensile strength of the remolded loess is related to consolidation pressure or compaction degree. The double electric layer overlaps and common water film is formed when adjacent clay particles are close to each other. The electrostatic force which is generated by the attractive effect of cations (between the soil particles) and negative charges (attached on the clay surface) increases the attractive force of soil particles.

Table 7.1 Calculation formulas of different tensile strength test.

Method	Formula
Uniaxial tensile	$\sigma_t = \dfrac{P}{\pi r^2}$ P: Destructive load r: Cylindrical sample's diameter
Three-point bending	$\sigma_t = 1.5\dfrac{PL}{Bh^2}$ B: Width (cm) h: Height (cm) L: Span (cm) P: Destructive load
Radial compression	$\sigma_t = \dfrac{2P}{\pi Ld}$ L: Sample's length d: Sample's diameter P: Destructive load
Axial compression	$\sigma_t = \dfrac{P}{\pi(bh - a^2)}$ a: Pad's diameter b: Cylindrical sample's diameter h: Sample's height P: Destructive load

With the increase of consolidation pressure and compaction degree of remolded loess, the internal porosity of loess is decreased, leading to the increase of dry density. Moreover, in consolidation, greater the number of layers indicates smaller tensile strength.

7.3.6 Test methods

Currently, the tensile strength of loess is usually measured by direct tensile methods or indirect tensile methods. The formulas are shown in Table 7.1.

Direct tensile methods are that the axial tension is directly applied to the sample to measure the tensile strength, including the uniaxial tension method and the three-axis tensile method.

Indirect tension methods are based on certain assumptions. The pressure or torque is applied to the specimen of loess until it is destroyed, and the tensile strength is calculated by the corresponding formula. Indirect tension methods include radial compression (Brazil splitting), axial compression, and three-point bending.

Under the same experimental conditions, the result of tensile strength measured by radial compression or axial compression is less than that measured by direct tensile methods for the same samples.

A linear relationship can be established between the test results of direct tensile and indirect tensile, and can be expressed as: $\sigma t.\ \text{direct} = k\sigma t.\ \text{indirect} + m$, and $R^2 > 0.9$ (Wang et al., 2015; Li et al., 2007).

As shown in Figure 7.7, the slope of the curves based on radial compression and axial compression is in the range of 1.034 to 1.527 (>1). This shows that the tensile strength measured by radial compression and axial compression test is less than that measured by direct tensile tests, and the value measured by radial compression is smaller compared with that of axial compression. The slope of the curve based on three-point bending is 0.4274 (<1), showing that the test result from three-point bending is larger than the others.

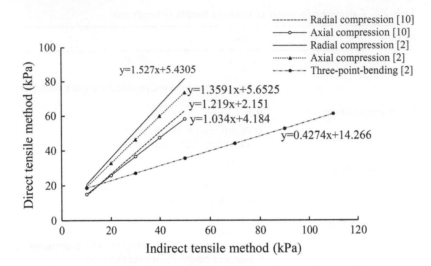

Figure 7.7 Relationship between direct tensile methods and indirect tensile methods.

REFERENCES

[1]   Dang, J.Q., Li, J. & Zhang, B.P. (2001) Uniaxial tension crack characteristics of loess (In Chinese). *Journal of Hydroelectric Engineering*, 4, 44–48.

[2]   Li, J.Y., He, C.R. & Tang, H. (2007) The comparative study on the tensile strength test on soft clay (In Chinese). *Subgrade Engineering*, 2007, (2), 104–105.

[3]   Li, R.J., Liu, J.D., Yan, R., Zheng, W. & Shao, S.J. (2014) Characteristics of structural loess strength and preliminary framework for joint strength formula. *Water Science and Engineering*, 7(3), 319–330.

[4]   Lu, Z.F., Zhao, Z.F., Cao, J.Y., Luo, H.F., Fan, J. & Chen, B.J. (2015) Experimental study on tensile strength of soil using uniaxial tension (In Chinese). *Geology of Anhui*, 51(2), 153–156.

[5]   Luo, Y.S. & Xing, Y.C. (1998) Tensile strength characteristics of loess (In Chinese). *Journal of Shaanxi Water power*, 14(4), 6–8.

[6]   Sun, M.X., Dang, J.Q. & Kang, S.X. (2006) Tensile character of disturbed loess (In Chinese). *Journal of Xi'an University of Arts & Science (Natural Science Edition)*, 9(3), 59–61.

[7]   Sun, P., Peng, J.B., Chen, L.W. & Wang, Z.X. (2009a) Experiment research on tensile fracture characteristics on loess (In Chinese). *Chinese Journal of Geotechnical Engineering*, 31(6), 980–984.

[8]   Sun, P., Peng, J.B., Chen, L.W., Yin, Y.P. & Wu, S.R. (2009b) Weak tensile characteristics of loess in China – An important reason for ground fissures. *Engineering Geology*, 108, 153–159.

[9]   Sun, P., Peng, J.B., Yin, Y.P. & Wu, S.R. (2010) Tensile test and simulation analysis of fracture process of loess (In Chinese). *Rock Mechanics*, 31(2), 633–637.

[10] Wang, Y.H., Ni, W.K. & Yuan, Z.H. (2015) Study on the test method for tensile strength of undisturbed loess (In Chinese). *Science Technology and Engineering*, 15(7), 1671–1815.

[11] Yuan, Z.H., Ni, W.K. & Wang, Y.H. (2015) Property of tensile strength of unsaturated loess (In Chinese). *South-north Water Diversion and Water Conservancy Science & Technology*, 13(1), 123–126.

Chapter 8

Loess geohazards in China

8.1 LOESS GEOHAZARDS IN CHINA

In China, loess is mainly distributed in the provinces along the middle reaches of the Yellow River. Loess covers an area of about 635,280 km² (Liu, 1965), accounting for 6.6% of China's land area and 4.9% of loess-covered areas in the world (Figs. 1.6 & 1.7). According to statistics, the population in loess-covered areas of China was approximately 293 million in 2000. This population is estimated to increase to 331 million by 2020, accounting for 22.8% of the total population in China (Peng et al., 2014). With the increase in the population in loess-covered areas and the rapid development of the economy, as well as the widespread engineering activities, the frequency of loess geohazards dramatically increased in recent years.

Geohazards in loess-covered areas of China mainly include collapse, landslide, earthflow, ground subsidence, and ground fissure. Among these geohazards, loess collapse and landslide are the most frequent and serious ones. On July 30, 2013, the loess collapse in Dudu Village in Linxian County of Shanxi Province resulted in three casualties (Fig. 8.1). On September 17, 2011, the loess landslide in Bailuyuan of

Figure 8.1 Loess collapse in Dudu Village of Linxian County, Shanxi (Lv, 2016).

Figure 8.2 Loess landslide in Bailuyuan, Shaanxi (Sun, 2013).

Shaanxi Province resulted in 32 casualties and serious disruption of traffic (Fig. 8.2). Loess geohazards have seriously threaten people's lives, properties, water resources, electric transmission, transportation and other infrastructure, which restrict the economic and social development in the Loess Plateau region. Therefore, the investigation and analysis of the influencing factors, distribution laws, development characteristics, and failure mechanisms of loess geohazards is necessary to prevent the occurrence of geohazards in loess-covered areas.

8.2 SHALLOW LOESS COLLAPSE

Loess collapse refers to the loess failure that occurs suddenly and violently in the shallow surface without distinct sliding surface, which has subclasses of peeling, sliding, toppling, falling, cracking–sliding and caving. The volume of the displaced mass is usually hundreds of cubic meters or less, and the displaced mass always accumulates at the slope toe.

8.2.1 Controlling factors

Joints and fissures

Loess slopes have many joints, of which the scales, origins, natures, and periods of formation are different (Figs. 8.3 & 8.4). Different joints intersect each other, creating a three-dimensional joint network. These structural planes are the channels for water infiltration and flow and form surfaces of soil for erosion, which also control separation. Loess joints are usually vertical, but have branches and turns. The joint branches are approximately parallel to the free loess surface, which often separate loess into loose blocks and create conditions for loess collapses.

Figure 8.3 Loess vertical joints (Linxian, Shanxi).

Figure 8.4 Loess joints and cracks (Linxian, Shanxi).

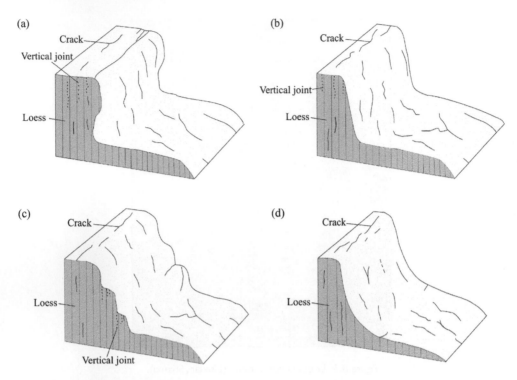

Figure 8.5 Four basic types of loess slopes: (a) convex slope; (b) rectilinear slope; (c) stepped slope; and (d) concave slope (Li et al., 2017).

Slope profiles in loess

The shape, gradient, and height of loess slopes are closely related to the occurrence of collapses. Previous studies generally divided loess slopes into four types according to their profiles, namely, convex, rectilinear, stepped, and concave (Fig. 8.5). Convex and rectilinear slopes are more prone to loess collapses because of abundant materials in the upper part (Figs. 8.5a and 8.5b). In addition, the shoulder of these slopes easily suffers from weathering compared with stepped or concave slopes (Zhu, 2014).

Taking the loess slopes in Yan'an City of Shaanxi Province as an example, the classification according to slope shapes shows that the number of stepped slopes is the largest, accounting for 40% of the total, followed by convex slopes, accounting for 30% of the total; rectilinear slopes, accounting for 20% of the total; and concave slopes, accounting for 10% of the total (Fig. 8.6) (Zhu, 2014).

Figure 8.7a shows the relationship between the slope shapes and the occurrences of loess collapses. Statistical analysis of 470 loess collapses in recent years in Yan'an City shows that the rectilinear slopes are the most prone to loess collapses compared with other types of slopes, accounting 45% of the total. Convex slopes account for 33% of the total. Stepped and concave slopes account for 11% of the total (Zhu, 2014).

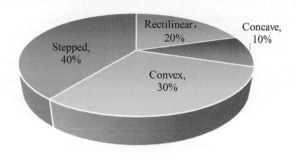

Figure 8.6 Distribution ratio of loess slopes with different shapes in Yan'an City (Data from Zhu, 2014).

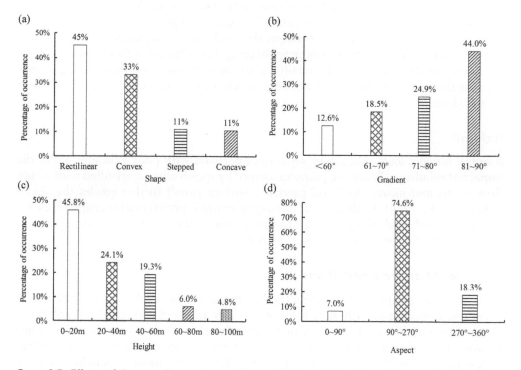

Figure 8.7 Effects of slope profiles on loess collapses: (a) shape; (b) gradient; (c) height; and (d) aspect (Zhu, 2014; Yang, 2010; Wei, 1995).

Figure 8.7b shows that failure occurs mostly on slopes with gradients greater than 60°, and the number of failures increases significantly with the gradient. The statistical analysis of the available data shows that 18.5% of the collapses occurred on slopes with gradients ranging from 61° to 70°, 24.9% of the collapses occurred on slopes with gradients ranging from 71° to 80°, and 44% of the collapses occurred on slopes with gradients ranging from 81° to 90° (Zhu, 2014). The gradient significantly affects the stress distribution inside the slope. The radial and tangential stresses transform into tensile stresses at the shoulder of the slope and develop a tension band. The steeper the slope is, the wider the tension band would be (Lei, 2001).

Figure 8.7c shows that slope height is another main factor that controls the occurrence of loess collapses. In Huangling County of Shaanxi Province, the collapses occurred on slopes with heights of 5 m to 60 m, accounting for 89.2% of the total number of occurrences. The remaining 10.8% occurred on slopes with heights of more than 60 m (Yang, 2010). A high slope normally develops a gentle gradient because of long-term weathering and erosion. By contrast, a low slope is generally steep (Chen, 2009), and thus, more prone to collapses.

Figure 8.7d shows that the south-facing slopes are more prone to the development of collapses than the shady slopes. The statistical analysis of 71 loess collapses in Yanchang County, Shaanxi Province shows that 74.6% of the collapses occurred on slopes with aspect angles ranging from 90° to 270°, particularly within 180° to 270° (Mao, 2008). This finding may be attributed to the fact that south-facing slopes receive long sunshine hours and soil temperature is relatively high during the day, resulting in large temperature differences between day and night. Furthermore, south-facing slopes are generally subjected to more weathering, resulting in fractures, which are not conducive to slope stability. Slope failures are more frequent on south-facing slopes because they are more populated and human activities cause widespread disturbances on the slopes.

Rainfall

Rainfall induces loess collapse by separating particles with poor adhesion under the impact of raindrops. When the potholes formed by splash erosion are filled with water, flow forms and moves small soil particles. Surface runoff further erodes the slopes (Zhang et al., 1994). In the case of persistent rainfall, preferential seepage pipes are usually formed inside the slope, saturating the soil, reducing the shear strength, and eventually leading to loess collapses (Fig. 8.8).

Activities of animals and plants

A negative correlation exists between the occurrences of loess collapses and vegetation coverage. Loess slopes with more trees and shrubs are more prone to loess collapses. During growth, the roots of trees or shrubs absorb moisture, introducing seepage pressure and softening the soil. Moreover, plant roots create cracks in the surrounding soil, which is called the root splitting effect (Fig. 8.9). Plant roots also benefit biological activities. The holes made by rats, snakes, and even worms work as the weak units in the loess body (Fig. 8.10) (Ye et al., 2013).

Human activities

Loess-covered areas of China have a population of approximately 300 million. Human activities are frequent and mainly involve cutting slopes for buildings, excavation for cave dwellings, construction of terraced fields, and construction of roads. Cutting slopes for buildings cause the side slope to become steep. Unloading-induced tensile fractures are usually produced on the trailing edge of the slope during rapid adjustment of the stress field within the slope (Fig. 8.11a). When a cave is excavated, roof damage

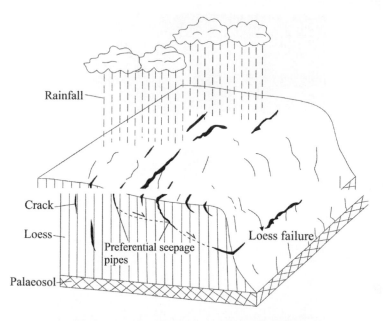

Figure 8.8 Loess collapse resulting from rainfall infiltration (Modified from Tang et al., 2013).

Figure 8.9 Root splitting effect (Xiangning, Shanxi).

Figure 8.10 Wormholes (Xiangning, Shanxi).

(normally caving) occurs because of the local tensile stress concentration when the design of the geometric section of the cave is improper (Fig. 8.11b). The terraced fields change the original path of the surface runoff and enhance rainfall infiltration. Together with irrigation, the terraced fields increase the water content of the loess slopes and the phreatic level (Fig. 8.11c). The majority of the traffic lines in loess-covered areas stretch along valleys and bank slopes. Slope cutting and excavation during road construction result in a large number of high and steep side slopes, which provide a breeding environment for failures (Fig. 8.11d).

Freezing and thawing

The freezing and thawing effects caused by seasonal transformation can destroy the structure of surface soil. In winter, with the decrease in ground temperature, the moisture at the toe of the slope does not easily evaporate. Moisture freezes, and then the soil is forced to expand, thus increasing porosity. In spring, ice in the pores gradually melts, and the pore structure is adjusted further.

Loess failures occur frequently from March to May. This period is the transition from winter to spring. The soil temperature rapidly increases from a value less than 0 to a value greater than 0. Figure 8.12 shows that the temperature of the soil remains negative and the frozen depth can reach approximately 1.0 m from December

Figure 8.11 Typical human engineering activities in loess-covered areas of China: (a) cutting slopes for buildings; (b) excavation for cave dwellings; (c) construction of terraced fields; and (d) construction of roads.

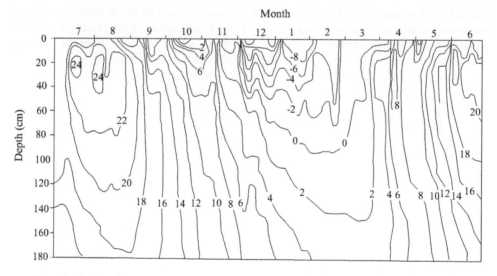

Figure 8.12 Monthly variation of ground temperature in loess slopes (Yang & Shao, 2000).

to February in loess-covered areas of China. At the end of March, the ground temperature begins to increase and the frozen layer gradually enters the thawing stage. The soil is rapidly heated up to approximately 8°C by mid-April.

Freezing and thawing mainly promote the occurrence of loess failures via the following three ways: (1) Frost heaving damages the soil structure and reduces soil shear

Figure 8.13 Loess collapse (Taiyuan, Shanxi).

strength. The loess itself contains a considerable number of large pores, and frost heaving further increases the distance between soil particles, reduces the dry density of soil, and loosens the structure, thereby reducing its cohesion and internal friction angle. (2) Thawing causes the loess to collapse and reduces its shear strength. Thawed water can dissolve the cement (particularly calcareous cement) between loess particles, damaging the loess structure and increasing pore water pressure, thereby reducing the shear strength of the soil (Pang, 1986). (3) Freezing and thawing accelerates the development of joints and fissures in the slope. Snow and ice melt as the temperature increases in spring, and meltwater seeps down to the joints and fissures. With the frequent day-and-night temperature fluctuations, the moisture cycles in the form of water and ice and facilitates the rapid modification of joints and cracks (Duan et al., 2012). The peeling failure that occurred on the road in Taiyuan (Fig. 8.13) is a result of the freezing and thawing effect.

8.2.2 Distribution law

Spatial distribution

(1) Loess strata

The three main layers of loess exposed in China are Wucheng loess (Q_1) deposited in the Early Pleistocene, Lishi loess (Q_2) deposited in the Middle Pleistocene, and Malan loess (Q_3) deposited in the Late Pleistocene (Tang, 2011). Lishi loess (Q_2) is reddish yellow clayey soil with layered calcareous concretion, characterized by dense structure and low water permeability. Malan loess (Q_3) is light yellow silt, with loose structure, large pores, dense vertical joints, and high permeability. Field investigations show that most of the loess collapses occurred in the Q_2 and Q_3 loess, particularly in the Q_2 layer.

(a)　　　　　　　　　　　　　(b)

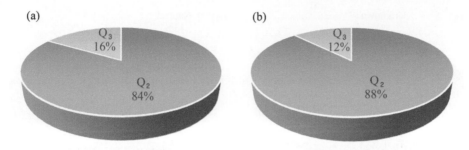

Figure 8.14 Collapses in different loess layers: (a) Shilou County; and (b) Xiaoyi County, Shanxi.

A total of 132 loess collapses occurred in Shilou County of Shanxi Province in the past 10 years, among which 102 loess collapses occurred in Lishi loess (Q_2) and 21 in Malan loess (Q_3). The remaining 1% occurred in Wucheng loess (Q_1) (Fig. 8.14). Records in Xiaoyi County, Shanxi for the same period shows that 88% loess collapses occurred in Lishi (Q_2) loess.

(2) River banks

Loess collapses often distribute densely along the valley on both sides of the river, and gully slopes are areas prone to loess collapses. The distribution density and the influence of the collapse are closely related to the development stage of rivers and valleys. In the early formation of the valley, vertical erosion by the river mainly occurs. Moreover, collapses occur frequently on both sides of the valley, but most of them are small. In the mature stage of the river, lateral erosion mainly occurs. Moreover, both sides of the river are strongly subjected to weathering, unloading, and corrosion, which easily lead to collapse. When the valley is in its topographic old age, human engineering activities increase in the wide valley, likely leading to the revival of old collapses and causing disasters.

(3) Human activities

Statistics show that more than half of the loess collapses were caused by human activities. This finding indicates that human activities have a significant influence on the stability of loess slopes. The more intense the engineering activities, the more frequent the collapses occur.

Cutting slopes for buildings, road construction, and excavation of cave houses are the main engineering actions in loess-covered areas. Cutting slopes will cause the concentration of stress in the toe of the slope. Then, tension cracks will form in the rear edge. The development of joints and fissures in loess easily causes loess collapses. If the design of the loess slope is unreasonable and the drainage and retaining measures are poor, then the accumulation effect caused by heavy mechanical vibration and the induction by rainfall can lead to the formation of loess collapses in the areas that the highway, railway, and other traffic lines pass through. If the lower part of the slope excavated by cave house dwellers is steep and erect, then the distribution of the bearing

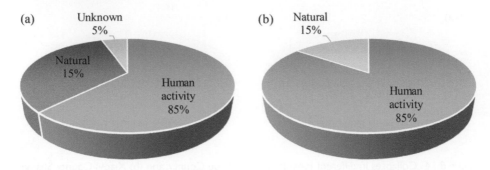

Figure 8.15   Contribution of human engineering activities to loess failures: (a) Shanxi; and (b) Huangling, Shaanxi (Data from Yang, 2010).

capacity and stress is concentrated, which will lead to a small kiln hole falling at the top of the loess cave house.

An investigation on loess failures, which has been conducted for five years in Shanxi Province and one year in Huangling County, Shaanxi Province, shows that more than half of the failures occurred because of human activities (Fig. 8.15). Among the 16 failure cases that occurred in 2014 in Yan'an, 9 failure cases were related to the over-steep slopes for the construction of cave dwellings and the other 7 failure cases were the consequence of the improper treatment of the side slopes for road construction (Lei, 2001). These findings indicate that the more intense the human activities are, the greater the probability of loess failures.

Temporal distribution

(1) The distribution law in a month

The seasonal variations of rainfall are significant in the Loess Plateau, although the annual average rainfall in this area is low. Rainfall is mainly concentrated from July to September, accounting for approximately 60% of the annual rainfall (Qian, 2011). Meanwhile, the rainfall in January and December account for only approximately 1% of the annual rainfall. Concentrated rainfall in the rainy season has become a major triggering factor of loess collapses. The freezing and thawing period is also prone to collapses. The soil freezes from November to December as the weather becomes cold. Then, the soil thaws with the increase in temperature from February to April. The contraction and expansion of the soil occur alternately, promoting the formation of collapse.

Figure 8.16 is a histogram of the monthly occurrence percentage of loess collapse in Shanxi Province from 2005 to 2010, Shaanxi Province from 1985 to 1994, and Gansu Province for many years. In Shanxi, most of the loess collapses occurred in July, followed by March and November, which are consistent with the law of rainfall and freezing. In Shaanxi, most of the loess collapses occurred in July and August by up to 46 times, accounting for 50% of the total, which have a direct relationship with heavy rain in the period. The loess collapse that occurred in Gansu Province concentrated in

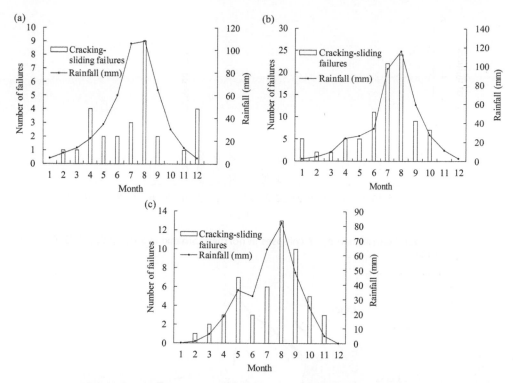

Figure 8.16 Relationship between loess failure and rainfall: (a) Shanxi; (b) Shaanxi; and (c) Gansu provinces.

seven months to nine months, with the most rainfall in one year. In addition, the month of May had relatively large rainfall, and thus, the number of landslides is relatively more than that of the remaining months.

(2) The distribution law in a day

Statistics show that loess collapse is highly likely to occur in the early hours of the morning. The statistical analysis of 32 loess collapse cases that caused deaths in the northern Shaanxi Province (Fig. 8.17) shows a high occurrence frequency of such failures between 9 pm to 4 am the next day, with the number of cases reaching up to 18 times or 56% of the total number of cases. The temperature difference between day and night is more obvious in the loess area than in other regions with the same latitude (Wang, 2004). Therefore, the slope expansion and contraction effect is obviously different in the morning and evening. With the effect of gravity, the slope constantly moves downward. With the cyclic action of shrinkage stress and expansion stress, the slope in the limit stable state becomes unstable. The loess collapse in Jiaokou Town, Yanchang County, Shaanxi Province, was exactly caused by diurnal temperature variation in winter (Mao, 2008).

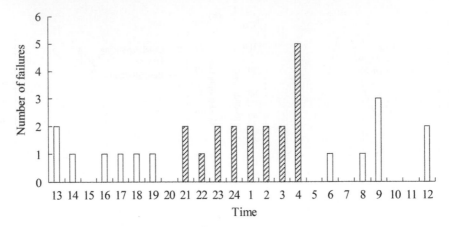

Figure 8.17 Temporal distribution of 32 collapses in northern Shaanxi in a day (Wei, 1995; Cai et al., 1988).

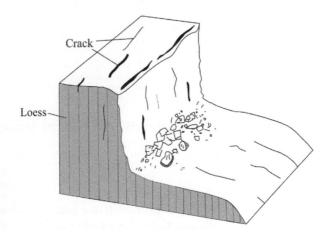

Figure 8.18 Peeling (Li et al., 2017).

8.2.3   Failure modes

Loess collapse is influenced by many factors, such as landform, vertical joints, meteorology, hydrology, and human activities; thus, different failure modes have been developed. In view of the mechanical characteristics of the deformation and instability of loess slopes, some scholars generalized six types of failure modes of loess collapse. These modes are as follows: peeling, sliding, toppling, falling, cracking–sliding, and caving.

(1) Peeling

Peeling collapse generally occurs in the foot and surface of the slope (Fig. 8.18). Peeling collapses occur because of several reasons. (a) The slope is strongly stimulated by

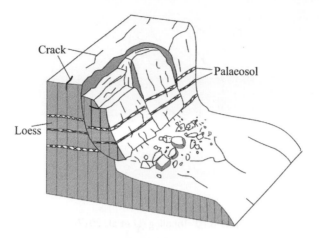

Figure 8.19 Sliding (Li et al., 2017).

rainfall and the erosion of surface water when rainfall is concentrated. (b) In the early stage of slope formation, the surface of the slope forms a hard shell and then spalls due to the change in water evaporation and temperature in soil. (c) In the external force of weathering, the loess layer with different compositions is destroyed along the bedding. The characteristics of peeling collapses are high frequency, small scale, and small damage depth. With the slope gradually eroded by peeling collapse, the impact of the collapse gradually expands, resulting in a significant loss of land resources.

(2) Sliding

After the shoulder of the loess slope is cut by vertical joints or cracks, the cut soil slides down the slope along the ancient soil layer or weathering surface. This phenomenon is usually referred to as a sliding collapse. In addition to gravity, the hydrostatic pressure and hydrodynamic pressure generated by rainfall soften the paleosoil layer or contact weathering surface, thus causing the soil to slide gradually. Once the center of gravity slides off the steep slope, collapse occurs (Fig. 8.19) (Tang, 2013).

(3) Toppling

Toppling collapse usually occurs in the loess layer with a soft upper portion and a hard lower portion. When the slope loses its stability, the collapse body topples at one point. The characteristics of this type of collapse mode are as follows. Toppling collapse involves many vertical joints and columnar joints on top of the slope, and the overlying loess is not suspended (Fig. 8.20). This collapse mode occurs through a variety of ways. (a) When the vertical loess slope forms a cavity due to long-term erosion, the supporting area of the loess block decreases, the gravity center of the loess body continues to shift, and the rear side cracks continue to expand. Under the action of gravity and bias stress, the vertical loess body forms a dump creep, and toppling collapse occurs. (b) When a special horizontal force (seismic force, hydrostatic pressure, hydrodynamic pressure, and frost heaving force) acts on the soil, the slope generates toppling deformation.

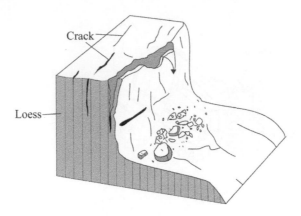

Figure 8.20 Toppling (Li et al., 2017).

(c) When a weak surface forms (from Li et al., 2017) at the foot, the water softens the slope, and then bias stress is produced, which can result in toppling collapse. (d) Vertical soil is bent due to the long-term action of gravity, which can also promote toppling collapses.

(4) Falling

Falling collapse often occurs in the part where the loess body is usually out of the slope and vertical joints are well developed in the trailing edge. With the force of gravity, the joints in the trailing edge of the protruding soil gradually expand. As a result, tension is concentrated on the parts that have yet to produce joints and fissures. Once the tensile stress exceeds the tensile strength of the soil, pull cracks rapidly develop downward until they completely pass through the prominent soil, which in turn collapses suddenly (Wang et al., 2011) (Fig. 8.21). In addition to gravity, a variety of factors such as shock, weathering, and plant root promote the occurrence of such collapses.

(5) Cracking–sliding

Cracking–sliding collapse occurs in the Q_3 loess with certain strength. Its characteristics include a steep slope and vertical fractures on top of the slope, although the surface structure, which is inclined, is not obvious (Peng et al., 2015). As a result of disturbances from human activities or when the toe of the steep slope is infiltrated by rain and the slope drainage is poor, the loess body falls down gradually along a plurality of parallel and tough surfaces, leading to cracking–sliding (Fig. 8.22), which is also known as a scattered collapse (Duan et al., 2012).

(6) Caving

Caving collapse usually develops in newly deposited loess or Q_3 loess (Peng et al., 2015). The excavation or disturbance processes at later stages easily cause the internal structure of loess slope unstable. Falling and collapse occur along the top of the cave,

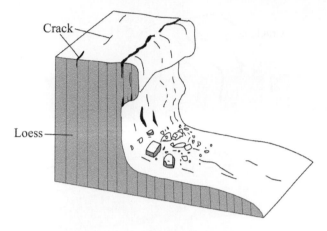

Figure 8.21 Falling (Li et al., 2017).

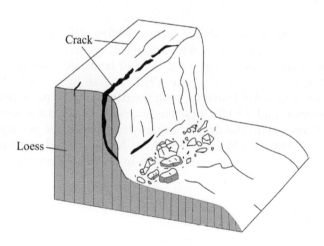

Figure 8.22 Cracking–sliding (Li et al., 2017).

which could eventually fail and break down. This phenomenon is commonly known as "the collapse of kiln". The profile of the collapse is usually a trapezoid (Fig. 8.23).

8.3 DEEP-SEATED LOESS LANDSLIDE

The term landslide is widely used to refer to the movement of a mass of rock, debris, or earth down a slope. In this work, landslide refers to the deep failure of slope with a distinct sliding surface.

Loess landslide is a phenomenon in which the loess mass is separated from a high and steep slope along a weak structural plane under the influence of its own gravity.

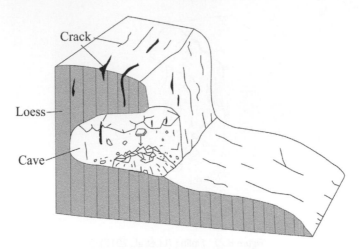

Figure 8.23 Caving (Li et al., 2017).

It is composed of landslide body, sliding surface, sliding wall, and landslide bed (Tang, 2013). China is one of the countries with a frequent occurrence of landslides (Luo & Wang 2001). Statistics show that about a third of landslide disasters in China occur in the Loess Plateau. From the 1950s to 1992, 16,616 loess landslides occurred in the northern part of Shaanxi Province, with the density being more than 5 landslides per square kilometer; and 14,109 loess landslides occurred in the eastern and western regions of Gansu, with the density being over 6 landslides per square kilometer (Lei, 2001). Since 2008, 1,131 loess landslides have occurred in Shaanxi Province; 1,300 in Lanzhou, Gansu Province; and 4,576 in the eastern part of Gansu Province (Peng et al., 2014).

8.3.1   Controlling factors

Slope shape

Loess slope is the basic geological condition for the formation and development of loess landslides. The characteristics of a slope landform determine the distribution of the internal stress and surface runoff of the slope. Particularly, slope geometry, height, and gradient, which are closely related to landslide events, are the main factors that determine the magnitude of sliding force (Fig. 8.24).

The slope profile is a prerequisite for landslide hazard. The geometric shape of a slope determines the size and distribution of stress in the slope, which controls the stability of the slope and the mode of deformation and failure. As shown in Figure 8.24a, a convex slope is the most prone to the development of landslide hazards, followed by a rectilinear slope and a concave slope (Chen, 2009).

Figure 8.24b shows that loess landslides occur mostly on slopes with heights ranging from 20 m to 40 m (Zhu, 2014; Yang, 2010). The stress state in the top, surface, toe, and bottom of a slope change with the slope height significantly, and eventually, this

Table 8.1 Failure modes of loess collapse.

Type	Main features					
	Properties of soil	Occurrence site	Landform characteristics	Shape of collapse body	Profile shape of failure surface	Stress state
Peeling	Occurs in newly accumulated loess or Q_3 loess	Toe or surface of slope	Weathering and erosion of slope is obvious.	Slice shape	Straight line type	Forces (weathering, erosion, disturbance)
Sliding	Occurs in newly accumulated loess or Q_3 loess	Surface or body of slope	Loess slope shows certain slope characteristics. A paleosol layer exists in the upper layers.	Comes in a variety of shapes, such as plates and wedges.	Polygonal shape, wedge shape, arc shape	Shear effect
Toppling	Occurs in the layer with a soft upper loess and hard lower loess	Top of slope	The slope is upright and steep, and vertical joints and column joints exist.	Plate shape, column shape	"L" type	Effect of overturning moment
Falling	Occurs in newly accumulated loess or Q_3 loess	Top of slope	The top portion develops weathering fracture and tension fracture, and the slope is vacant with a prominent loess body.	Block shape	Straight line type	Gravity effect
Cracking–sliding	Occurs in newly accumulated loess or Q_3 loess	Top or surface of slope	Steep slope, vertical cracks, no tendency of free face surface structure	Plate shape, column shape	Ladder type	Shear caused by gravity
Caving	Occurs in newly accumulated loess or Q_3 loess	Inside of slope body	Human engineering activities exist in the slope.	Funnel shape	Trapezoid	Gravity effect

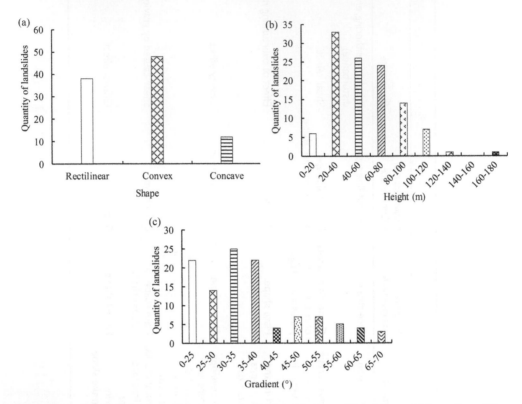

Figure 8.24 Effect of slope profiles on loess landslides: (a) shape; (b) height; and (c) gradient. (Zhu, 2014; Yang, 2010; Wang, 2013).

phenomenon causes different parts of the slope to undergo deformation and destruction (Zhang et al., 1994). Under the same conditions, the safety factor of a slope decreases gradually with an increase in slope height (Chen, 2003).

Loess landslides usually occur on slopes with gradients within 40°. As shown in Figure 8.24c, slopes with gradients in the 25°–40° range are most prone to landslides, and slopes with gradients below 25° are not as prone to landslides.

Human engineering activities

With the rapid economic development in the Loess Plateau in recent years, the number of construction projects that involve excavating loess slopes has increased because of limited topographical conditions. As a result, loess landslides have increased in number. The loess landslide that occurred in Zaolin, Shanxi, on October 7, 2009 was caused by human engineering excavation, which destroyed the stress equilibrium state of the original slope; this landslide resulted in six deaths (Fig. 8.25). On March 12, 2010, the loess landslide in Zizhou County of Shaanxi Province occurred as a result of the construction of houses, which disturbed the excavation slope stability, as well as freezing-and-thawing conditions; the landslide resulted in 27 deaths (Fig. 8.26).

Figure 8.25 Landslide in Zaolin, Liulin, Shanxi (Tang, 2013).

Figure 8.26 Landslide in Zizhou, Shaanxi (Tang, 2013).

Earthquakes

Earthquakes could cause loess landslides (Fig. 8.27), and their number gradually declines from high intensity regions to low intensity areas. The impact of earthquakes on loess slopes is described as follows: (a) earthquakes cause direct damage to soil structure and decreases the inherent coupling force in soil particles; (b) earthquakes cause the liquefaction of fine sand strata and saturated loess layers; (c) earthquakes increase the falling power of slopes. For instance, the Wenchuan Ms 8.0 earthquake triggered a large number of loess landslide disasters in the Loess Plateau in 2008.

Rainfall

The rainfall in the Loess Plateau of China is concentrated, although the annual average rainfall in this area is low and the erosive precipitation is primarily rainstorm and

Figure 8.27　Shuiwan landslide caused by an earthquake (Li et al., 2016).

persistent rainfall (Ye, 2012). Most loess landslides occur during the rainy season in areas with high rainfall. Loess stability is influenced by the infiltration of rainfall, although the permeability of loess is low. At present, field observations after rainfall and artificial simulated rainfall infiltration experiments all indicated that the depth of rainfall infiltration is limited. Nevertheless, a long period of rain is known to trigger a number of small shallow loess landslides. Shallow loess landslides are caused by the direct infiltration of rainfall to the surface of loess, which leads to the softening of the surface layer and an increase in weight. Rainfall is known to be concentrated underground through channels, causing groundwater levels to rise and perched water to form; the result is generally a large-scale loess landslide (Li & Zhang, 2011).

Freezing and thawing

The Loess Plateau of China has a semi-arid continental monsoon climate. It is cold and dry in winter, and the temperature in spring rapidly rises. The temperature difference between day and night is large, and seasonal freezing and thawing is dominant. The temperature interannual variability in the loess slope zone indicates that (Fig. 8.12) seasonal freezing and thawing play a major role in the loess area. The slope surface begins to freeze from October to November and then thaws from February to April every year. The minimum temperature can reach $-30°C$, and the maximum seasonal frozen depth is from 1.0 m to 2.5 m. The region is prone to not only seasonal freezing and thawing but also landslides. When the groundwater is uneven, forms vein structures in the slope, and then discharges in spring, the effect of seasonal freezing and thawing can promote the occurrence of landslide events, including the enrichment and expansion of groundwater in the slope, increase of slope soil softening and weight, and rise in static and dynamic water pressure; all these conditions decrease the stability of the slope and lead to landslides.

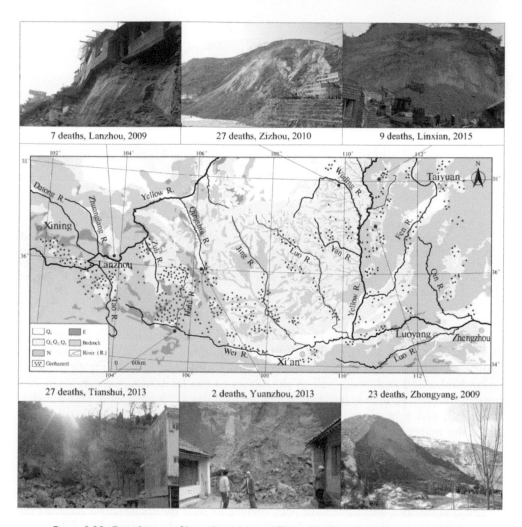

Figure 8.28 Distribution of loess landslides in China (Modified from Peng, et al., 2014).

8.3.2 Distribution law

Spatial distribution

The general spatial distribution law of loess landslides lies in a "group". That is, regardless of the size of the area, the "group" of loess landslides is always obvious (Fig. 8.28).

(1) Law of regional distribution

According to their regional distribution, loess landslides in China are bounded roughly by a 400 mm rainfall isoline (i.e., Lanzhou, Xiji, Huanxian, Jingbian County, Yulin). The distribution obviously covers the north to south zones. Landslides are highly concentrated in the south, gradually becoming sparse in the north.

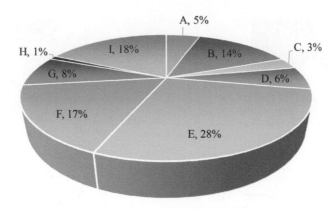

Figure 8.29   Distribution of loess landslides in Yan'an City (Wang, 2013).
  Legend: Landform units: A – Loess-covered middle mountain; B – Loess platform with gullies; C – Loess residual platform with gullies; D – Loess platform ridge with gullies; E – Long ridge-shaped loess hill with gullies; F – Ridge hillock-shaped loess hill with gullies; G – Hillock ridge-shaped loess hill with gullies; H – Gentle ridge and broad valley-shaped loess hill; I – Terrace.

Each large cluster of loess landslide distribution is controlled by a combination of slip factors, such as regional landform, stratigraphic structure, hydrological and hydrogeological conditions, and erosion intensity and depth. According to the scale of landslides and regions as well as the causes of the landslides, we can divide the combination of slip factors into five groups, namely, basin, downstream, plateau edge, structural, and urban types (Jin & Zhang; 1996). The edge of the basin, the slope along the margin of the basin, the fault zone, and the surrounding rock are often the dominant distribution points of loess landslides. As shown in Figure 8.27, the NS distribution of the strip cluster on the right side is located in the Yellow River Valley and Shanxi Lvliang Mountain. At the lower middle part from the east of Baoji, the linear distribution in the north shore of the Weihe River includes more than 180 large old landslides, and some have been revived or near the resurrection state. At the left-middle part, a cluster of loess landslides represents the magnitude 8.5 Haiyuan Earthquake of 1920 in the west of the Liupan Mountain.

(2) Uneven distribution in different landform units

Statistics show that in the landform units of loess platform with gullies, long ridge-shaped loess hill with gullies, and terrace, loess landslides occur frequently. By contrast, in the landform units of loess residual platform with gullies, platform ridge with gullies, and gentle ridge and broad valley-shaped loess hill, the frequency of loess landslides is correspondingly low (Fig. 8.29). This characteristic may be related to the evolution stages of different landform units.

(3) Concentrated development in areas with intense human engineering activities

In the Loess Plateau of China, cutting slopes and digging caves have long been practiced. Although the economy has improved, cutting slopes to construct houses and kiln

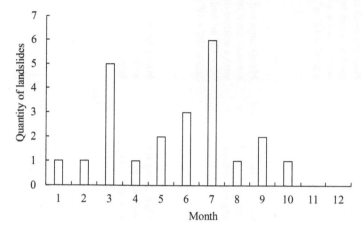

Figure 8.30 Relationship between the number of loess landslides and months in the Heifangtai platform, Yongjing, Gansu (Wang, 2004).

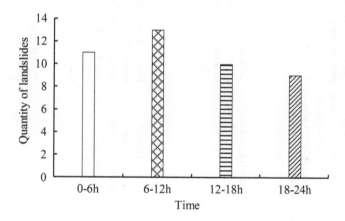

Figure 8.31 Time distribution of 44 loess landslides in the period of one day (Mao, 2008).

hoops and build engineering facilities according to the restrictions of terrain conditions is inevitable. If the original balance of the loess is broken by cutting slopes, loading, and other functions, unloading, tension, and weathering cracks occur in the loess slope. During the rainy season, landslides become increasingly frequent. In addition, irrigation, construction of reservoirs and highways, and other engineering economic activities disrupt the original stress balance in slopes, leading to slope instability and landslides.

Investigations into the distribution of loess landslides in Yan'an City in recent years show that the number of landslides is higher in residential areas than in other places, thus indicating the landslides are closely related to unreasonable engineering activities.

Table 8.2 Failure modes of loess landslides (according to geological structure) (Chen & Shi, 2006).

Landslide type	Section sketch map	Common landform	Sliding surface position	Sliding surface shape	Scale
Homogeneous loess landslide		Slopes with large angles and thick loess layers in the edges of loess platforms and loess tablelands, loess ridges, and hillocks.	Top of paleosoil with perched water.	Approximately circular, relatively smooth. Rear controlled by a vertical joint. Relatively steep.	Most are small and medium-sized landslides with a volume of less than 10×10^4 m^3.
Loess–bedrock interface landslide		Loess hilly areas.	Contact surface of loess and mudstone, sandstone, or shale of Tertiary system, Trias, or Jurassic.	With a relatively flat and rectilinear gradient. Dip angle of mostly 10°–20°.	Mainly large and medium-sized landslides with a volume of over 10×10^4 m^3 and, in some cases, 100×10^4 m^3.
Loess–bedrock bedding landslide		Edges of loess platforms and loess tablelands, loess ridges, and hillocks.	Main sliding surface: outward inclined weak layers and interlayers in the bedrock. Sliding surface in upper loess: secondary cracking surface.	Main sliding surface with flat and rectilinear gradient controlled by rock stratum occurrence. General dip angle between 10° and 20°.	Mainly large and super large landslides.
Loess–bedrock cutting layer landslide			Main sliding surface: cutting rock layer along the joints, fissures and, other structural surfaces.	Main sliding surface with relatively steep rear.	

Temporal distribution

In different months of the year in a specific area, the frequency of the development of loess landslides is different under the influence of rainfall and the freezing and thawing effect, which shows some regularity. However, at different times of the day, the occurrence of loess landslides does not show any regularity.

(1) The distribution law in a month

In the Loess Plateau, large-scale landslides occur in two peak periods in a year: one peak period occurs in the rainy season, and the other is concentrated in the freezing-and-thawing season in early spring (Luo, 2010) (Fig. 8.30).

Most loess landslides occur in the rainy season spanning six to seven months. In 1984, Tianshui experienced abundant rain all year round. Rainfall soared in July, causing a large number of loess landslides, including over 80 landslides in Tianshui County, 220 landslides in Xihe County, and 180 landslides in Lixian. since the Ming Dynasty, 73% of landslides took place in the months of July, August, and September, and only 15% occurred from March to May in the Linxia area, Gansu Province.

In the loess region of Northwest China, the effect of seasonal freezing and thawing on slope stability is significant.

The freezing and thawing of the Loess Plateau occurs from February to April. This period is also the peak period of loess landslides, with only a few landslides occurring in other periods (Ye, 2012). Since the 1980s, a series of large-scale loess landslides have occurred, and they include those that took place in Saleshan, Jiang Liu, Gu Liu, Tianshui, Huangci, and so on. These loess landslides cause serious significant economic losses and casualties.

(2) Distribution law in a day

Statistics show that the regularity of loess landslides occurring at a certain period in one day cannot be ascertained (Fig. 8.31). Affected by heavy rains, engineering activities, and earthquakes, the distribution of loess landslides in a certain period within a day could vary.

8.3.3 Failure modes

In terms of landslide occurrence conditions, causes of formation, thickness, damage mechanism, and characteristics of motion, the failure of loess landslides can be divided into three modes: homogeneous loess landslides, loess–bedrock interface landslides, and loess–bedrock landslides (Table 8.2). This classification scheme considers the development position of the sliding surface and its relationship with the rock strata. Such scheme is conducive to the identification and prevention of landslides in loess areas.

REFERENCES

Cai, B., Chen, B., Wei, L.M. & Xiong, D.K. (1988) Preliminary study on development law of landslide and landslide in the Three Gorges Reservoir Area (In Chinese). *Bulletin of Soil and Water Conservation*, 8(2), 18–24.

Chen, R.B. (2009) *Study on the survey method and laws of geological disaster in Loess Plateau-A case study of Wuqi County, Shaanxi Province* (In Chinese). Msc Thesis. Chang'an University, Xi'an.

Chen, Y.M. & Shi, Y.C. (2006) Basic Characteristics of Seismic Landslides in Loess Area of Northwest China (In Chinese). *Journal of Seismological Research*, 29(3), 276–280.

Chen, Z.Y. (2003) *Stability analysis of soil slope* (In Chinese). Beijing, China Water Conservancy and Hydropower Publishing House.

Duan, Z., Zhao, F.S. & Chen, X.J. (2012) Types and Influencing Factors of Collapse Development in Loess Plateau Region of North Shaanxi: A Case Study of Wuqi County (In Chinese). *Journal of Natural Disasters*, 21(6), 142–149.

Jin, Z.X. & Zhang, S.W. (1996) Factors of landslide hazards and their application (In Chinese). *Gansu Journal of Science*, 8(Supp.), 123–128.

Lei, X.Y. (2001) *Geo-hazards and Human Activity at Loess Plateau* (In Chinese). Beijing, Geological Press.

Li, H., Yang, W.M., Huang, X., Liu, T., Tian, Y. & Cheng, X.J. (2016) Characteristics and deformation mechanism of Shuiwan seismic loess landslide in Maiji, Tianshui (In Chinese). *Journal of Geomechanics*, 1(22), 12–24.

Li, T.L. & Zhang, M.S. (2011) Triggering factors and forming mechanism of loess landslides (In Chinese). *Journal of Engineering Geology*, 19(4), 530–540.

Li, Y.R, Mo, P. & Xu, Q. (2017) Classification of loess failures. J Asian Earth Sci (under review).

Liu, T.S. (1965) *The Deposition of Loess in China* (In Chinese). Beijing, Science Press.

Luo, D.H. (2010) *Experiment for loess landside of freezing-thawing* (In Chinese). Msc Thesis. Xi'an University of Science and Technology, Xi'an.

Luo, Y.S. & Wang, G.L. (2001) *Research and Engineering of Collapsible Loess* (In Chinese). Beijing, China Building Industry Press.

Lv, M. (2016) *The present situation of the loess disasters in Shanxi Province and the collapse of geological water sensitivity analysis* (In Chinese). Msc Thesis. Taiyuan University of Technology, Taiyuan.

Mao, S.L. (2008) The Yanchang County geology disaster distributes the rule and the stability analyses (In Chinese). Msc Thesis. Chang'an University, Xi'an.

Pang, G.L. (1986) A Discussion on Maximum Seasonal Frost Depth of Ground (In Chinese). *Journal of Glaciology and Geocryology*, 8(3), 253–254.

Peng, J., Li, X.Y. & Yan, R.X. (2015) Failure modes classification and countermeasures of loess collapse in Northern Shaanxi area (In Chinese). *Journal of Yangtze River Scientific Research Institute*, 32(10), 11–16.

Peng, J.B., Lin, H.Z., Wang, Q.Y., Zhuang, J.Q., Cheng, Y.X. & Zhu, X.H. (2014) The Critical Issues and Creative Concepts in Mitigation Gation Research of Loess Geological Hazards (In Chinese). *Journal of Engineering Geology*, 22(4), 684–691.

Qian, P. (2011) *Study of types for highway drainage system in loess areas in Northern Shaanxi Province* (In Chinese). Msc Thesis. Chang'an University, Xi'an.

Sun, W.Q. (2013) *Research on mechanical mechanism of excavation and unloading landslide for the example of the landslide of Baqiao* (In Chinese). Msc Thesis. Chang'an University, Xi'an.

Tang, D.Q. (2011) Factors for geological disaster about loess landslide and landslip and distribution regularity (In Chinese). *Journal of Xiangfan University*, 32(11), 49–53.

Tang, D.Q. (2013) *Study on the Loess landslide Mechanism of Slope Toe Excavation* (In Chinese). PhD Thesis. Chang'an University, Xi'an.

Tang, Y.M., Xue, Q., Bi, J.B., Sun, P.P. & Cheng, X.J. (2013) Preliminary study on loess landslide rainfall triggering models and thresholds (In Chinese). *Geological Review*, 59(1), 97–106.

Wang, B. (2013) *The correlation reseach on geomorphology and geological disasters in Yan'an* (In Chinese). Msc Thesis. Chang'an University, Xi'an.

Wang, G.L., Zhang, M.S. & Su, T.M. (2011) Collapse Failure Modes and DEM Numerical Simulation for Loess Slopes (In Chinese). *Journal of Engineering Geology*, 19(4), 541–549.

Wang, N.Q. (2004) *Study on the growing laws and controlling measures for loess landslide* (In Chinese). PhD Thesis. Chengdu University of Technology, Chengdu.

Wei, Q.K. (1995) Collapse Hazards and Its Distribution Features of Time and Space in Shaanxi Province (In Chinese). *Journal of Catastrophology*, 10 (4), 55–59.

Yang, S.H. (2010) *Study on the growing characteristics and rules of the geological hazards on Loess Plateau-Taking Huangling County as an example* (In Chinese). Msc Thesis. Chang'an University, Xi'an.

Yang, W.Y. & Shao, M.A. (2000) *Study on Soil Moisture in Loess Plateau* (In Chinese). Beijing, Science Press.

Ye, W.J. (2012) Test research on mechanism of freezing and thawing cycle resulting in loess slope spalling hazards in Luochuan (In Chinese). *Chinese Journal of Rock Mechanics and Engineering*, 1(31), 199–205.

Ye, W.J., Wang, P. & Yang, G.S. (2013) Formative factors of loess collapse and method for determining its influence range (In Chinese). *Journal of Engineering Geology*, 21(6), 920–925.

Zhang, Z.Y., Wang, S.Q. & Wang, L.S. (1994) *Engineering Geological Analysis Principle* (In Chinese). Beijing, Geological Press.

Zhu, J.H. (2014) *Study on The relationship between slope geometrical morphology and landslide collapse disasters in Yan'an* (In Chinese). Msc Thesis. Chang'an University, Xi'an.

Index

Printed and bound by CPI Group (UK) Ltd, Croydon, CR0 4YY

01/11/2024

01782603-0002